Geography of the 'Ne
Education Market

Secondary school choice in England and Wales

CHRIS TAYLOR
Cardiff University School of Social Sciences

Routledge
Taylor & Francis Group

LONDON AND NEW YORK

First published 2002 by Ashgate Publishing

Reissued 2018 by Routledge
2 Park Square, Milton Park, Abingdon, Oxon OX14 4RN
711 Third Avenue, New York, NY 10017, USA

Routledge is an imprint of the Taylor & Francis Group, an informa business

Copyright © Chris Taylor 2002

The author has asserted his moral right under the Copyright, Designs and Patents Act, 1988.
to be identified as the author of this work.

All rights reserved. No part of this book may be reprinted or reproduced or utilised in any
form or by any electronic, mechanical, or other means, now known or hereafter invented,
including photocopying and recording, or in any information storage or retrieval system,
without permission in writing from the publishers.

Notice:
Product or corporate names may be trademarks or registered trademarks, and are used only
for identification and explanation without intent to infringe.

Publisher's Note
The publisher has gone to great lengths to ensure the quality of this reprint but points out
that some imperfections in the original copies may be apparent.

Disclaimer
The publisher has made every effort to trace copyright holders and welcomes
correspondence from those they have been unable to contact.

A Library of Congress record exists under LC control number: 2001053656

ISBN 13: 978-1-138-73455-5 (hbk)
ISBN 13: 978-1-138-73440-1 (pbk)
ISBN 13: 978-1-315-18726-6 (ebk)

Contents

List of Figures

List of Tables

Preface

Since 1979 there has been a marked shift in the education system of England and Wales, and, in particular, in the provision and organisation of compulsory schooling. One of the key components of this shift was the introduction of Open Enrolment, which gave parents the opportunity to state a preference over the school they would like their children to attend. This study examines the secondary education system and specifically focuses on issues of equity in the 'new' education market, both in the process of parents choosing a school and the product, or outcome, of this new system on school admissions. This is done from a geographical perspective, and consequently makes comparison between different Local Education Authorities (LEAs) and different schools, urban and rural. Using Geographical Information Systems this study examines patterns of competition and choice based on pupil home postcodes and relates these patterns to the decision-making process of parents. This book presents the geography of the 'new' secondary education market and provides a conceptual framework that stresses the importance of the geographical context behind competition and choice. This research also shows that consideration of 'local' markets is necessary in aiding an understanding of the reforms, and that the outcome of competition between schools tends to reflect their relative examination performances. However, it is also clear that parents from different socio-economic backgrounds are 'active' in the 'new' education market, which, consequently, has in the majority of cases prevented further social segregation of intakes, and has in some cases actually reduced social polarisation. There is a cautionary note to these findings since the study also shows that there are a small number of schools, which due to their extreme levels of popularity and unpopularity, have seen the socio-economic composition of their intakes change dramatically, increasing the social divide between these schools.

Acknowledgements

I wish to thank the LEAs, schools and parents who provided information and access when they were consistently under pressure to 'perform'. To me this research will have failed if the significance of the findings and conclusions of this study do not match the energy and time of those individuals who helped me.

I would like to thank everyone who has made the completion of this book possible including the following, Ken Fogelman, Michael Bradford, Tom Whiteside, John Fitz and Stephen Gorard. Particular thanks must go to Patrick Bailey for his enthusiasm early on, to Gareth Lewis for his great encouragement and support, and to many friends and colleagues at the University of Leicester Geography Department.

Material from Crown Copyright records has been made available through the Post Office, ESRC Data Archive and the Ordnance Survey.

1 Introduction

Introduction

Since 1979 there has been a marked shift in the education system of England and Wales, and, in particular, in the provision and organisation of compulsory schooling. The most influential changes were the Conservative Government's 1988 Education Reform Act and the succeeding series of Acts, Charters and White Papers which created a form of 'market' system for the delivery of education and the organisation of provision. One of the key components of this shift was the introduction of Open Enrolment, which gave parents 'the right to express a preference as to where they would like their children to go to school' (Department for Education, 1992, p.28). Very quickly this was trumpeted by Government Ministers as 'parental choice', choice implying empowerment to the individual, when in fact the policy was simply a desire to improve education standards and simultaneously reduce the number of surplus places in the system. The arrival of a new Labour Government in 1997 has done very little to alter the course of change and has indeed fully supported the continuation of 'parental choice'.

Programs of school choice have been introduced in many countries such as in the US (Cookson, 1994), New Zealand (Lauder and Hughes, 1999), The Netherlands (Dronkers, 1995), Belgium (Vandenberghe, 1999), Sweden (Lidström, 1999), Canada (Manzer, 1994) and Scotland (Adler *et al.*, 1989). While differences may exist in the extent marketisation of public education has occurred in those countries (Whitty, 1997; Teelken, 1999), it still remains that England and Wales provides a set of reforms on one of the largest scales. 'Parental choice' has been introduced across the whole of England and Wales at both the primary and secondary level, alongside a number of key mechanisms meant to install a market-based system of delivery.

Such transformations of education provision parallel significant shifts in the provision of other public welfare services, such as housing, health care, social care and pensions, which are increasingly dependent upon private resources and market mechanisms. Politically, the New Right has argued that neo-liberal policies are necessary to keep pace with new demands upon the Welfare State. But, as a result there is a general concern that such an

1

approach to welfare provision is socially divisive, thus creating a climate such that 'the issue of the influence of markets on public sector welfare provision and in education in particular is one of the most important research topics of the 1990s' (Bowe *et al.*, 1992, p.24).

The objectives of the 'new' education market are now being achieved by making schools more accountable as well as increasing competition between them by allowing parents to indirectly choose the 'good' schools and weed out the 'poor' ones. In effect this has involved the creation of a quasi-market (Bartlett, 1993) within which parents and children are the consumers, the schools the producers, and the commodity being exchanged is the education. In other words, there has been a 'shift away from collective-welfare orientation ... towards an individual-client orientation' (Adler, 1993, p.2).

This commodification of education has drastically altered the nature of compulsory education, particularly secondary education. The reforms to the organisation of schooling have given parents greater responsibility to ensure that their children receive the highest levels of education in particular schools. Consequently, the stakes have increased creating situations, reported in the broadsheet newspapers, of 'The Great School Lottery' (The Independent, 1997a, p.14), 'School squeeze leaves children without places' (The Independent, 1997b, p.15), and 'The admissions nightmare' (The Independent on Sunday, 1997, Which School? p.7), and which has led to the formation of new 'tactics' to get the school of your choice, 'Bending the school rules' (The Sunday Telegraph, 1996, p.18), 'Tricks to get into the right school' (Daily Telegraph, 1996, p.14) and 'Parents who move house to be top of the form' (Weekend Telegraph, 1996, p.16). The media's portrayal of choice in education has always been careful to select examples, and areas, on which to base their stories. However, our understanding of the 'new' education market is still rather limited and, typically, confined to the boroughs of London.

Within academia the central focus has been on Open Enrolment and both the process of school choice and the outcome of such decisions (see for example Ball *et al.*, 1995; Glatter *et al.*, 1997b; Gorard and Fitz, 1998). The key issues have been whether parents have real choice in the 'new' education market, and if so, whether all parents have this choice, how the decision-making process for school choice operates and how it varies between parents, and what the consequences are for schools and their intakes. The greatest debate has been as to whether the education market is equitable, or, specifically, if 'cream-skimming' (Le Grand and Bartlett, 1993) within the market has arisen. In Whitty's own review of research on school choice in England, the USA and New Zealand, 'the academically able are the 'cream' that most schools do seek to attract. Such students stay in the system longer and thus bring in more money, as well as making the school appear successful in terms of its test scores and hence attractive to other desirable clients' (1997, p.14). Detailed research by Ball (1993) and Gewirtz *et al.* (1994) have shown how different parents perceive and engage with the market in a variety of ways, which in turn produces different outcomes reflecting social class divisions. However, more recently, research examining the 'new' education market across England and Wales has suggested that the market, in fact, is creating more homogenous school intakes, i.e. greater integration of pupils from different socio-economic backgrounds (Gorard, 1998; Gorard and Fitz, 2000).

These two, very different, perspectives of equity, or fairness, in the market place reflect the *process-side* and the *product-side* of the market. Even though the research appears to illustrate two different scenarios there is every possibility that they could exist alongside each other. A crucial element to research in this field is to combine these two elements of equity together, and then to decide if equity in one, and inequity in another, might still produce a socially 'just' system of education provision (Smith, 1994; Hay, 1995).

Another important feature of these two research perspectives is their methodological distinction. On the one hand, *process-side* research tends to be small-scale, ethnographic and limited to a small number of Local Education Authorities (LEAs). But, on the other hand, almost conversely, *product-side* research has been large-scale, empirical and across many different LEAs. Once again, to further an understanding of 'parental choice' in the 'new' education market, both, the decision-making process, and the final choice of schools, requires research to consider the market at many spatial scales in order to link together the findings of the *process-side* and the *product-side* of the market. A geographical approach to such research provides the means to consider the many spatial scales of the education market.

A Geographical Perspective and Research Methodology

The geography of education has been a relatively underdeveloped area of research (Bradford, 1990) but clearly has a very important role in any understanding of variations in education provision and performance across space, and in particular the dimensions of school choice and competition. For example, Bradford (1990) has illustrated the place that the geography of education has within geography (Figure 1.1), bringing together key components of economic geography, social geography and political geography.

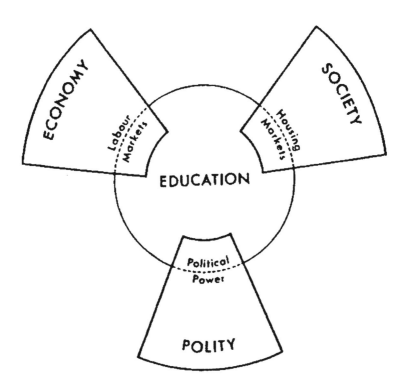

Figure 1.1   The place of the 'geography of education' in geography

Source: Bradford (1990, p.4)

Within research into the 'new' education market there have been many attempts to consider a geographical perspective (see Adler *et al.*, 1989; Gewirtz *et al.*, 1994; Glatter and Woods, 1994) but few have primarily set about their research with this as their main objective. For example, Gewirtz *et al.* (1994) discussed the importance of context and considered a socio-spatial model of analysis as a way of conceptualising the hidden constraints upon the market. However, their methodology did not incorporate the geography of the education market at the outset, and consequently, has never fully considered the education market in rural or even suburban locations. The general inclination of such research has been to incorporate geography as a way of understanding and organising the findings. This has clearly been useful, and this research argues that the geography of the education market provides the ideal means of conceptualising the processes and patterns that emerge. However, it is also necessary, in order to produce a full geographical appreciation of the market, to undertake a geographical methodological framework to which the processes of choice and competition, and the varying constraints that underpin these processes can be considered at several spatial scales, ranging from the Local Authority level down to the individual household.[1]

This research was carried out in England between 1995 and 1997, a period when the 1988 reforms were well established but finishing just before the introduction of a new Labour Government. The majority of research in England and Wales on school choice occurred at the beginning of the reforms and hence, it could be argued, only represented the initial start-up period of an education market. Gorard (1997), on the other hand, chose to examine the fee-paying sector in education in order to provide evidence of an 'established' education market.

As this is a geographical examination the analysis presented here focuses on three different spatial scales of the education market: the LEA, the education market place, and the consumers (i.e. parents). Consequently the research is based on three different sources of information, respectively. Much of the analysis considering the context to the reforms – the 'geography of education' – uses data published by the Department for Education and Employment at the level of the LEA.

The research examining patterns of competition and choice at the scale of the education market place is unique and innovative. Primarily it uses the home postcodes of all pupils transferring in the 1995/96 academic year between primary and secondary school in eight LEAs. These postcodes are introduced into a Geographical Information System (GIS) in order to undertake a spatial analysis of movement of pupils in the market place. The eight LEAs are representative of areas in England with very different geographies, such as London Boroughs, the deindustrialised Metropolitan Boroughs of the north and the west midlands, more affluent suburban counties, and more rural isolated county authorities. These areas of detailed study presented here provide the most extensive coverage, to date, of the impact of market reforms on schools in England and Wales. For instance, most research on 'parental choice' has only been conducted in London Boroughs, arguably very different to other areas of England and Wales.

The third main source of data provides information at the level of the consumers. This is based on a large-scale survey of parents whose children transferred to secondary school also in the 1995-96 academic year. As the focus of enquiry gets more detailed then so does the geographical coverage. The parents' survey focuses on the decision-making process of more than 200 parents with children at eight schools from the LEA sample outlined above. The schools were selected based on the market characteristics that they exhibited in the analysis of competition derived from the pupil postcodes.

The methodological framework for this research was carefully constructed in order to provide as comprehensive and significant a review of the impacts of market-based reforms to secondary schooling as possible. It ensures that comparisons can be made between areas while maintaining a good deal of detail in the analysis. Similarly, it tries to encompass the impacts of the market reforms at many spatial scales, such as at the level of institutions, the level of market competition, and at the level of the individual in the market place.[2]

Aims and Objectives

The primary aim of this research, therefore, was to develop a link between geography and educational research with particular reference to the 'new' education market and 'parental choice' at the secondary school level. It also aimed, from a geographical perspective, to detail the principal components of the education market that aid and constrain parents' choices of schools, by critically discussing the many education markets that exist. The third main aim was to focus upon the social equity and 'cream-skimming' debate that is currently of great importance to the future of these reforms.

Within these three broad aims there were a number of objectives that this research attempted to achieve:

- to conceptualise the spatial form of the education market;
- to appreciate regional variations within the education market, particularly across urban and rural areas;
- to understand how different education markets in different Local Authorities influence choice and competition of schools;
- to determine the different forms of competition which exist, and produce a general classification of such competition;
- to outline and interpret the decision-making process for choosing a school;
- to appreciate the spatial and social constraints upon the decision-making process;

- to explore spatial and social inequalities on access to choice and in the final choice of schools;
- to identify the degree of 'cream-skimming' and social inequity of the market place in the final allocation of pupils to schools; and
- to consider future policy implications as a result of these findings.

A Conceptualisation of the Education Market

Having outlined the aims and objectives of this research the discussion will now presents a conceptualisation of the education market. The analysis that precedes this introduction is organised around this. This conceptual framework was based upon a preliminary observation of the distribution of intakes and patterns of competition, and based upon previous research into the education market for secondary schools in a semi-urban East Midlands county. It provides a useful spatial tool from which to examine the processes and dynamics of the 'new' market mechanisms in education, and also provides a framework that combines *diversity*, *differentiation* and *hierarchy*, key elements of the English and Welsh education market (see Glatter *et al.*, 1997, for a discussion of these).

The basis of this conceptual framework is organised around the notion of the **market place**. The market could be considered as the abstract mechanism for allocating supply and demand, but the market place was defined as the physical locale of where most schools competed with each other, and where the majority of children attended schools. Thus the market place is the spatial arena in which the education system currently operates. Conceptually, there are four components to the market place, each associated with the four key actors and agencies used to organise the methodological framework, outlined earlier. These four components of the market place are:

i.   Institution Space
ii.   Producers
iii. Consumers
iv.   Competition Space

The first three components require further elaboration before outlining how all four components relate to the market place.

Institution Space

The **institution space** is the bureaucratically defined area in which education has traditionally been administered, the LEA. The consideration of the institution space goes beyond the administrative processes that operate, such as the way LEAs decide to administer admissions to schools, or the services that LEAs provide to schools. There are two important features of the institution spaces that help to characterise the market place; the historical development of the organisation of education and the geographical nature of the institution space. Both of these are discussed in greater depth in Chapter 3, but the decisions and responses of the LEAs have helped to determine the constraints and opportunities for the education market to develop.

Producers

Identifying the producer of secondary education in the quasi-market is not straightforward. But in the context of the market place, schools, and schools alone, are seen as the producers. They are the organisations that are producing the final commodity that is being 'purchased' by the 'consumer'. They have also been given the responsibility to promote their product within the market place and how to target their product to the available consumers. This responsibility has had to also fit around the introduction of the National Curriculum. However, it can easily be argued that there are particular school characteristics that differentiate them, such as the overall examination performance of their pupils, or the reputation the school has with the local population.

Most of the decisions that a school takes come from the management team within a school; the headteacher, senior staff and the governors. These key actors are not necessarily experienced in operating within a more commercial environment and therefore approach the market place in varying ways. As a result, schools approach their roles of producer and competitor in different ways. Some schools have attempted to avoid competition by creating agreements with neighbouring schools. One such consortium has included four secondary schools in rural Devon (The Observer, 1997).

At the centre of understanding the marketisation of schools and how well they perform within the market place is the spatial location of these producers. There are at least five ways in which the actual location of a school plays a significant role in the way producers operate and perform in the market place:

i. Proximity to other schools – characteristics of these adjacent schools will affect their own marketing.
ii. 'Community' role of school – schools could use this to market a 'local' education that meets the needs and wants of their locale.
iii. Traditional 'local' intake – can determine past examination performance of school and reputation of school.
iv. Environmental and aesthetic conditions of school location – can make a school (un)attractive to consumers.
v. Accessibility of school to consumers – this includes provision of school transport by the LEA and the provision of public transport.

Consumers

The 'consumers' within the new education market are parents and their children. The choices made by the parents and their children have to be seen in the context of potential constraints that they face within the education market place. There has been a great deal of research into the social constraints on such decisions (Ball *et al.*, 1995), such as the impact upon differing levels of empowerment and cultural capital. These two factors are unevenly distributed across the market place and can often define the final outcomes within the education market.

The location of the consumers is not just important in determining the access to alternative schools, it is also important in the acquisition of cultural capital required to make certain choices. Other consumers in the local neighbourhood can help provide the necessary information to know how to go about choosing a school and to produce a good deal of information about the different schools. This collective information can help to produce more informed decisions in making a school preference, such as 'hot' knowledge from social grapevines.

However, the major constraint on school choice is the amount of available choice, largely determined by the location of the families to the schools. This overlaps considerably with the discussion in the previous section on producers. The proximity to schools can often determine the amount of knowledge of those schools, particularly if they are community schools that organise activities for the local community. The location of the consumers can also affect the amount of accessibility they have to the schools. Are they located on an organised or public bus route? How long would the journey take to a particular school? These questions can prove to be crucial in the decision-making process. However, overriding even these decisions is the availability of pupil places in popular schools.

Chapter 6 takes a closer look at the consumption of schools and education from the perspective of the 'consumers', but the discussion now draws together these three components of the market place to construct a conceptual framework for analysing the 'new' education market.

The Market Place

The discussion now brings these three components together to form the conceptual framework, or market place, for this 'geographic' analysis of market mechanisms in the education system of England and Wales. Figure 1.2 illustrates a three-dimensional view of this market place.

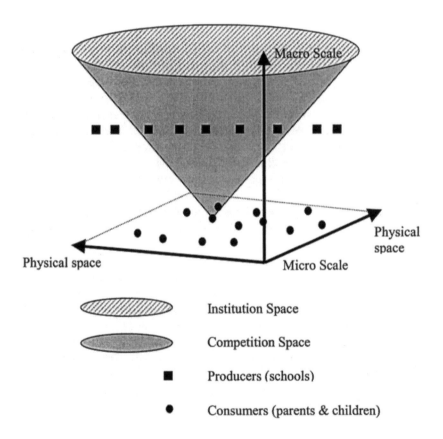

Figure 1.2   Conceptualisation of the market place

From this diagram it is clear that both producers, i.e. schools, and consumers, parents and their children, are located in physical space and interact with the market place at different scales. For example, parents engage with the market at the level of individual decision-making, the more micro-scale of the market place. This differs from schools, where their decisions and actions are responses from, or targeted to a set, or collection, of individual consumers. This more macro-scale perspective by schools is clearly related to the spatial structure of the market place. However, this relationship varies between producers. The most obvious distinctions are between LEA-maintained primary and secondary schools. Primary schools tend to engage with the market with fewer individual consumers in mind to secondary schools. Consequently, their relationship with the locale tends to be stronger, i.e. these schools are more embedded in the social spaces of smaller communities. The ultimate macro-scale phenomenon of the market place is generally the LEA,[3] who has, over time, defined and organised the **institution space**. This often defines the spatial extent of the market place, but is no way a well cast blueprint for the market place.

A basic parameter of the conceptual framework illustrated in Figure 1.2 is that all parents and children, the consumers, are treated as equal, except for their location in space, which acts as a constraint upon their decision-making. Therefore, based on their proximity to schools, consumers view an area from which they can choose a school. The resulting area, defined by the relationship between the location of consumers and schools, creates a **competition space** in which the majority of market activity takes place.

The notion of competition spaces has often been used in studies of retail markets and (Marsden *et al.*, 1998; Hughes, 1999), in particular, the organisational activity of the supplier-retailer-purchaser relationships in space. In the new education system similar relationships have emerged. Figure 1.2 represents an example of one consumer's view of the schools and the subsequent competition space that ensues. The competition space, therefore, defines the spatial arena in which schools are likely to compete with one another. Since the competition space, in this conceptualisation of the market place, is defined by the location of schools and consumers, some schools are located in areas where there will inevitably be greater demand from parents and consequently greater intensity of competition between producers.

This is further illustrated in Figure 1.3, which gives a two-dimensional example of the varying intensities of competition as defined by the locations of three individual consumers, all things being equal.

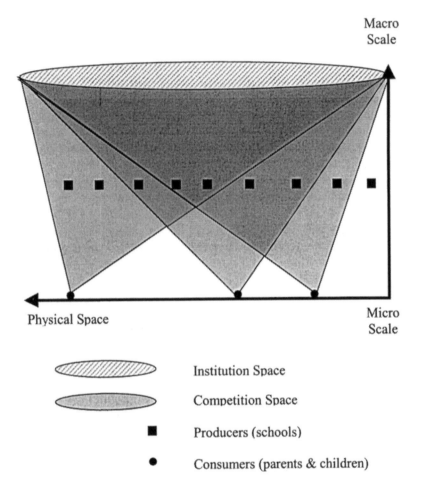

Macro
Scale

Physical Space

Micro
Scale

Institution Space

Competition Space

■ Producers (schools)

● Consumers (parents & children)

Figure 1.3 Two-dimensional view of the market place from the perspective of three consumers illustrating the varying intensities of competition for schools

This conceptualisation of the market place provides a basis, or framework from which to examine the reality of producer competition and consumer choices. For example, from observing the intake patterns for many schools it soon becomes clear that there are great variations in the spatial extent of intakes.

This would suggest that, again assuming all parents are equal except for their locations, secondary schools engage with the market place at different

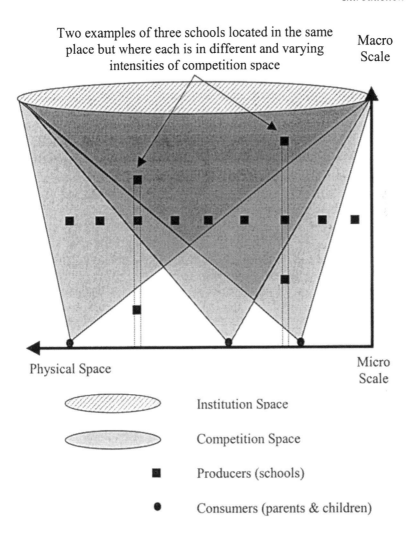

Two examples of three schools located in the same place but where each is in different and varying intensities of competition space

Macro Scale

Physical Space

Micro Scale

Institution Space

Competition Space

■ Producers (schools)

● Consumers (parents & children)

Figure 1.4   **Two-dimensional view of the market place with two situations showing the relationship between market place scale and competition spaces**

scales. This can be seen in Figure 1.4, which shows a two-dimensional view of the market place and the resulting competition spaces from the perspective of three consumers. In this case there are two examples given, in which three schools are located in the same place but where each is engaging with the market place at different scales. Consequently, the competition spaces and intensity of competition they face are different.

There could have been many factors that position them at different scales in the market place, such as whether they are academically selective schools, religious schools or how good their examination performances are. But, whatever reason this might be, the framework argues that these schools are likely to have different relationships with their locale and with the consumers.

This conceptual framework only provides a hypothetical example of how the market in education operates. However, it provides an organisational and interpretative framework from which **market places**, **competition spaces**, **producers** and **consumers** can be compared, in order to understand the many other factors that influence and constrain competition and choice. This is very important, since the central issue to this study was the social equity of the market in education provision, and, hence, sets up the basis for interpreting social inequalities, both, in the choice process and as a consequence of the market mechanisms. The framework simply levels out the 'playing field' in which market mechanisms work, but provides the means to interpret and understand the processes and dynamics of the education market.

Outline of Book

This research is built around a simplified model of the process of choosing a school in the 'new' education market (Figure 1.5), which highlights three key areas of the process – (I) the choice of schools being made, (II) the context in which the decisions are being made, and (III), the final outcomes of who goes where.

However, before presenting the findings of the research Chapter 2 outlines the background to the research. In this Chapter the broader political context to recent reforms, both, across the Welfare State, and education specifically, are considered. In particular, it focuses on the development of a quasi-market approach to provision as well as the legislative changes involved before discussing recent critiques of these changes. The four subsequent Chapters examine the education market at four different spatial scales (the LEA, the market place, the school and the parents), each representing a shift down in the scale of the market place as well as covering the three areas of 'parental choice' outlined in Figure 1.5. Chapter 3 begins by considering the importance of the Local Education Authority (LEA) in defining the area in which most choice and competition takes place, and the influence it has on the education market. The Chapter then compares LEAs according to their diversity, or private-state continuum, and the levels of market activity and choice by parents. From this discussion it is possible to suggest that there are many different education markets, of different sizes, operating differently, and with varying levels of school choice available.

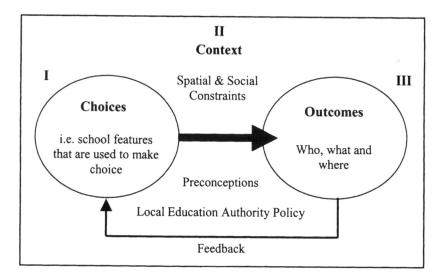

Figure 1.5   A simple model of 'parental choice'

The fourth Chapter extends this discussion of the LEA by examining competition between schools within a number of Local Authorities. Using admissions data for all of the secondary schools in these LEAs, patterns of competition are identified and conceptualised before attempting to classify different forms of competition, and identifying their unique market, educational and geographical characteristics. Again, this Chapter highlights the differences between LEAs and between schools, but also stresses the similarity of competition that the quasi-market is producing. Chapter 5 provides a link between Chapter 4 and Chapter 6 by outlining the school and intake characteristics of eight sample schools and the form of competition between them, thus, providing the specific context to understanding the decision-making process of school choice discussed in Chapter 6. In Chapter 6 the focus is upon the way parents went about choosing a school for their child, the stages in this process, how they acquired knowledge of the schools, how they made their final decision of school, and how the education market might be affecting the relationship between schools and their locale. Chapter 7 then draws together the findings presented in the previous four Chapters and considers the social impacts of the education reforms. It does this by examining the potential social inequities of both the *process-side*, i.e. which parents are more 'active' in the market, and the *product-side*, i.e. which pupils are in which schools of the education market, by returning to the admissions data for the sample of LEAs. The *product-side* of the market is divided into two parts: between-school social segregation across each LEA, and social polarisation of particular school intakes. The rather mixed findings of equity in the market, uncovered in Chapter 7, are returned to in Chapter 8, where they are discussed in terms of whether the market is socially just, and the implications of these findings on the future of education provision specifically and the quasi-markets in general.

Notes

1 The rationale behind this study is explored further in Chapter 2.
2 For a greater presentation of the methodological framework and methods employed throughout this research see Taylor (2000).
3 Clearly the LEA does not provide the definitive boundary to the market place to some parents. This is discussed in more depth in Chapter 3, however, this framework could allow for overlapping LEAs.

2 Quasi-markets and Educational Reform

Introduction

The transformation of education provision over the last two decades in England and Wales, and to some extent also in Scotland, is intrinsically tied to changes in the political economy of not just the UK but also the Western World. A changing global economy, the rise of the New Right in policy decision-making and a 'tired' Welfare State has prompted significant changes in policy across many spheres of public services, not least in education. One of the key concepts developed to reflect these changes has been that of **quasi-markets**. The notion of market principles while maintaining some form of state control have typically been applied to different areas of public provision and provides a useful theoretical tool to review legislative changes and evaluate the implications of the new policy strategy. The purpose of this Chapter is, therefore, to consider the development of a quasi-market in education alongside the broader changes to the political economy of the Welfare State as a basis for the rationale of this study.

This broad aim is broken down into four key objectives. The first is to outline the changes in the political economy of the British Welfare State that underlie the introduction of new reforms into public services generally, by chronicling the 'crisis' in the Welfare State and the subsequent rise of the privatisation debate heralded by the New Right. The many reasons for this shift in policy orientation are presented here in order to help understand the justification for educational reform. The second objective of this Chapter is to introduce the idea of market models of provision and the theoretical development of quasi-markets. These were initiated by a change in policy strategy seen under the Thatcher and Reagan administrations in the UK and USA respectively, and point to two common features, the introduction of market principles and consumer choice. The discussion then draws on these two as the basis for the emergence of quasi-markets. The third feature to be discussed in this Chapter follows on from the theoretical development of the quasi-market by considering the legislative changes in education provision since 1979 and highlights how these transformations in policy are closely related to the quasi-market model for the provision of public services. The final objective of the Chapter is to then critically review and evaluate the transformation of the organising principles of school provision. Using this review of the literature and the writings on policy research the Chapter concludes with outlining the rationale for this study.

The Welfare State and Policy Reforms

The Welfare State in the UK has been evolving since the end of the 19th Century but the most significant development came after the Second World War at a time when there was a general feeling that the State should be instrumental in providing security for society. The Beveridge Report of 1942 outlined how this could be achieved and called for the official introduction of the Welfare State. Governmental White Papers during 1944 eventually led to a series of legislative reforms between 1945 and 1948 that set up the Welfare State that was to remain relatively unchanged for the next 35 years. By the end of the period of legislative reforms the Welfare State featured several key elements (Bruce, 1968):

i.   Security of minimum incomes.
ii. Protection against accidents of life via insurance.
iii. Provision of special protection for children via family allowances.
iv. Universal provision for education and health care.
v.   Provision of environmental and welfare services, such as housing and children's welfare.

Beyond these specific elements, Papadakis and Taylor-Gooby (1987) outlined four 'systematic' roles that the Welfare State undertook. According to the authors, at a basic level it provided a system for meeting minimal needs; economically it assisted with regeneration and maintenance of the economic fabric; politically it became an important part of Government policies to gain popularity with the electorate; and, finally, at a general level it provided a useful means to bind society together.

The development of the Welfare State after 1945 has traditionally been considered within a political economy perspective, or as a 'double settlement'. The 'economic settlement' has involved the management of the economy in an attempt to remove or reduce natural oscillations in the economy. On the other hand, the 'political settlement' sustained these reforms by means of bi-partisan political support from both the Labour

Party and the Conservative Party. However, Clarke (1996) identifies two further settlements that have been important to the development of the Welfare State; the 'social settlement', which saw the rhetoric of the Welfare State being described around the idea of 'nation, family and work', and the 'organisational settlement' based around bureaucratic administration and professionalism. To Clarke (1996) the 'organisational settlement' has been influential in sustaining the Welfare State between the 1940s and the 1970s. Both bureaucratic administration, promising 'social impartiality', and professionalism, promising 'application of valued knowledge', have, alongside the bi-partisan 'political settlement', created what Clarke has called a 'triple social "neutrality"' capable of giving the Welfare State momentum for nearly half a century (Clarke, 1996, p.69).

However, during the 1970s these 'settlements' became unstable and, in particular, the important 'organisational settlement' began to be perceived as an enemy of society (Clarke, 1996). Proponents of change believed that the Welfare State was in 'crisis', politically (Johnson, 1991), economically and morally (Davies, 1991). From another perspective the functions of the Welfare State were also criticised as producing male advantage, enabling dominant ethnic groups to suppress minority groups and maintaining social class cleavages (Papadakis and Taylor-Gooby, 1987).

In terms of the political economy, however, there were deeper causes for change as a result of the ineffectiveness of Keynesian economic management. Taylor-Gooby (1998) suggests that the three most significant areas of change influencing policy direction were:

i.   Changing international political economy.
ii.   New styles of economic organisation.
iii.   New patterns of social need.

The first of these, the changing international political economy, involved an expanding market in currency exchange, greater international trade and the rise of multi-national corporations, which, it is argued, resulted in a reduction in the powers of the State. New styles of economic organisation were seen, predominantly in the private sector, as a shift from Fordist to post-Fordist means of production (Jessop, 1994), while the new patterns of social need arose from an ageing population and rising levels of unemployment, putting extra pressures on the Welfare State. Specifically to the UK, these changes were the backdrop to the oil-crisis of the mid-1970s and negative economic growth in the 1980s. Both of these events contributed to fiscal pressure on the Welfare State and the subsequent reduction of the State's role.

However, as Gamble (1991) notes, the pressures for change were not unique to the UK and the USA, yet changes to the Welfare State in these countries contrasts quite dramatically with many other countries. Ball (1998) goes further, and adds three caveats to the 'Fordist/post-Fordist and global economy' explanation for change. The first was that the Fordist means of production was not simply replaced but 'exported' to developing countries as part of a growing global economy. Secondly, the role of flexible specialisation in post-Fordist production was not necessarily comprehensively introduced to all sectors of the economy, particularly for service sector jobs. The third caveat comes from the work of Harvey who said that Fordist production and post-Fordist production are 'not alternative forms of capital and regulation but a complex of oppositions expressive of the cultural contradictions of capitalism' (Harvey, 1989, p.39).

As Whitty and Edwards (1998) discuss, an alternative explanation for changes in the Welfare State is necessary and hence they concentrated on the policy networks that have emerged in the last twenty years, particularly across the Atlantic between the UK and the USA. These policy networks can be seen in the context of another fundamental cause of change in approach to welfare services, the rise of the New Right with its associated neo-liberal ideologies. This is most clearly evident in the successful election to power of the Thatcher and Reagan administrations, in the UK and the USA respectively, in the 1970s. One of the most influential parts of the New Right policy was the introduction of 'monetarist' economic policy as a means of controlling inflation.

With this key approach to the economy the New Right provided the vehicle to change in the Welfare State that was seen as necessary alongside these greater changes in the political economy.

The basic objective of the new Conservative Government in 1979, what became known as Thatcherism, was to 'roll back the state' (Hudson, 1992) and reduce the involvement of the state in the provision of welfare. Dunsire (1990) identified four routes taken by the Conservative Government to meet this objective:

i.   Liberalisation or Conservative free-market ideology.
ii.   Debureaucratisation.
iii. Privatisation power.
iv.   Increased public choice.

Similarly Taylor-Gooby (1998) identified three key elements to the Thatcher reform programme; 'retrenchment', or the need to reduce costs (see Pierson, 1994), 'consumer choice', based on Hayek's (1960) normative theory of individual freedom, and 'labour market flexibility'.

The success of these new approaches to policy is discussed later but they do illustrate the context to the particular changes in the Welfare State and provision of public services. Most importantly they begin to outline the reasons behind the search for new forms of provision of these services. The most notable of which was the rise of the 'privatisation' debate involving a shift from the public sphere of provision to the private sphere of provision. This change in approach is closely related to a general blurring of the public and private sectors, both in production and consumption (Dicken and Lloyd, 1981).

Privatisation of the school system can be conceptualised by three areas of private influence; ownership, finance and control (Peston, 1984) (Figure 2.1). This is a useful illustration of the privatisation process since the *degree* of privatisation can be seen as being dependent upon the *degree* of private influence in these three areas and the impact on three different aspects of school provision; school levels, school types and examining bodies (Figure 2.1). Mulford (1996) has highlighted in detial the ways in which education in the UK has been privatised.

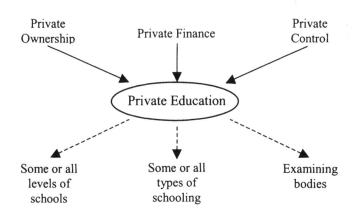

Figure 2.1 Conceptualising the privatisation of the school system

Source: Based on Peston (1984)

However, it would be too simplistic to conceive that the only approach to a change in the provision of state services was just 'privatisation'. Two other key concepts in understanding the shift in policy are 'managerialism' and 'consumerism' (Clarke, 1996). Ideologically, as Clarke outlines, managerialism represents 'good business practice' by learning from private business in order to make for a more 'dynamic, competitive, efficient and customer centred' Welfare State (1996, p.74). Practically, this meant that the welfare agencies would acquire new operating disciplines. The third key concept in understanding social policy reforms was that of 'consumerism'. Clarke argued that this came later in the Conservative Government's terms in office and was based on the rhetoric of the New Right's critique of the Welfare State that suggested the State was 'insensitive, undynamic and unresponsive' (1996, p.74).

These three concepts considered together determined a new strategy for providing welfare services, sometimes known as 'public sector managerialism' (Cutler and Waine, 1997), and have become accepted territory for subsequent Governments to the Thatcher administration, John Major's Conservative Government (1990-97) and Tony Blair's Labour Government (1997 to present). According to Papadakis and Taylor-Gooby (1987) the introduction of private welfare has remained seriously unchallenged because of two developments. The first has been the shift in social attitudes away from collectivism and towards individualism, seen in many different aspects of life. The second is at an academic level and which incorporates a new consensus centred round dissatisfaction with state welfare.

Consequently, public sector managerialism has become a very important feature in the provision of public services and the resulting debate on privatisation has led to change in policy, a reduction in state involvement and the encouragement of private sector in what was traditionally 'state' welfare.

The Market Model for Provision and Quasi-markets

Many of the changes to public services have been part of a 'hollowing out' (Jessop 1994) of the public sector generally. Pinch (1997) acknowledges the different methods used to bring about change in the provision of welfare services: Investment and technical change; Intensification; Flexibilisation; Contracting-out; Internal markets; Commercialisation and corporatisation; Devolution; and Decentralisation. Several of these are based on a 'market model' of provision. This means that they are based around market principles, involving supply, demand, some form of price, and consumer choice. The 'market model' has been particularly influential in steering reforms to the UK educational system.

According to Le Grand and Robinson (1984) there are two reasons as to why a 'market model' could be perceived as to be the most effective means of determining the quality and form of a service and who the service is for. First of all it gives the consumer free choice, without constraint, which in turn leads to the second reason that what is required will be provided. However, the same authors note five problems with the 'market model' which are summarised as:

i.   Capital market imperfections.
ii.   Imperfect information.
iii. Externalities.
iv.   Monopoly.
v.   Equity considerations.

Le Grand and Robinson (1984) conclude that a market system of provision is unlikely to meet its own objectives and that, therefore, Government intervention and regulation is required. However, some see the problems, or imperfections, of the market as providing the essence of effective policy (for example, Barry, 1991).

According to Taylor-Gooby (1998) offering choice in the provision of welfare services has to be viewed within the context of greater economic choice that individuals were expressing during the 'eighties' and 'nineties'. Taylor-Gooby (1998) also suggests that three factors played a significant role in this greater economic choice. The first has been the shift in the circumstances of individual life, particularly because of greater disposable incomes and greater personal savings. Secondly, wider economic and social changes have prompted such economic choices to be of personal significance, and thirdly, the ideological shift at the political level towards a neo-liberal approach, which uses the language of individual and free choice in all sectors.

In order to understand in more depth how and why a 'market model' can provide welfare services it is necessary to examine the theoretical conceptualisation of both the market principles and consumer choice that underpin this approach.

Theoretically, the two aspects of market principles and choice within the 'market model' of provision are relatively underdeveloped. This is particularly true for the theoretical consideration of choice, which has rarely gone beyond a debate on the economic motivation or instrumental rationality that is often assumed to drive choice within the market model.

As Taylor-Gooby (1998) highlights, this is often an oversimplified view of choice in provision and consequently calls for a more behavioural approach when considering choice as a mechanism for provision.

The most advanced theoretical consideration of the market model, and the use of market principles to organise and provide public services, has been that of Julian Le Grand and the concept of the **quasi-market** (Le Grand and Bartlett, 1993; Bartlett *et al.*, 1994). It should be noted that the concept of the quasi-market is very much a *product* of policy changes rather than a pre-determined approach to policy.

The characteristics of quasi-markets are that they are 'markets' 'because they replace monopolistic state providers with competitive independent ones', and are 'quasi' 'because they differ from markets in a number of key ways' (Le Grand and Bartlett, 1993, p.10). They differ because they do not necessarily seek to maximise profits or are not privately owned, consumer purchasing power is not expressed in monetary terms and it is not necessarily the direct user who exercises choice (Le Grand and Bartlett, 1993). Levacic (1995) believes the key to identifying a quasi-market is if the role of the Government or State shifts from being a 'provider' of services to simply the 'purchaser' of the same services. Cutler and Waine (1997) support this definition because they have identified it as a common feature of quasi-markets for different services.

Using the conceptualisation of privatisation, as illustrated in Figure 2.1, it also becomes clear that a quasi-market can be regarded as a product of some degree of privatisation in one or more of the key areas of ownership, finance and control. Consequently the quasi-market can be seen as a hybrid version of a totally 'private' system. Quasi-markets are meant to replicate the virtues of the market via the processes of competition. Such competition is then meant to produce a number of outcomes, such as greater efficiency in the provision of the service, decentralisation of authority in the provision, enhanced consumer choice delivering what is wanted and in the form required, diversity of providers intended to satisfy many and varied demands, and to produce a high quality service or product (Cutler and Waine, 1991). But the service is still publicly financed and subject to government intervention and regulation.

It soon becomes clear that quasi-markets meet the requirements of a neo-liberal ideology towards public services and in fact represent many of the elements of reform in all sectors of the Welfare State. However, the quasi-market as a concept is not totally accepted by all policy makers. For example, Thomas (1996) has highlighted the many definitions of the 'market' used in theoretical discussions of welfare reforms. Instead of an unclear quasi-market, Thomas stressed the importance of exchange relations under different influences as an alternative conceptualisation for changes in provision based around two concepts, interest and decision.

Using Thomas' framework, the form of 'allocative mechanism' in the reforms will exist between four 'ideal' types: command; market; collective; or college systems. However, the terms, objectives and processes evident in policy reforms are clearly related to principles of the 'market', in whatever hybrid form they appear. Consequently, the quasi-market allows these principles to be used as a framework and an analogy for studying the changes in the provision of welfare services, and in particular reforms to the educational system. This argument is strong enough to support the claim that in 1988 'the Government began to apply a programme of introducing internal or 'quasi'-markets to the Welfare State' (Le Grand and Bartlett, 1993, p.2). This began with the decentralisation of decision-making and the introduction of competition in provision. The first set of reforms that clearly reflected a quasi-market approach to provision was in compulsory education with the passing of the 1988 Education Reform Act. Other areas of public service soon followed with legislation reflecting the concept of the quasi-market. The shift in the role of the State from 'provider' to 'purchaser' was evident in the Local Government Act (1988), the Further and Higher Education Act (1988) and the National Health Service and Community Care Act (1990).

There has been much research into the impact of the quasi-market in these other public service sectors, but the educational system provides a unique example of the quasi-market in which the role of the consumer is much stronger and related to individual decisions. It is also significant because the 'consumer' and the 'user' are the same. The next section examines the reforms to education specifically by considering the different pieces of legislation introduced during the last 20 years.

Legislative Change in Education, England and Wales

The introduction of educational legislation and the subsequent changes over time can be seen in three waves.[1] Brown (1990) suggests that the first wave was the development of mass schooling during the 19th Century and the first half of the 20th Century culminating in the 1944 Education Act. This was followed by the second wave of legislation that involved an ideological shift in the organising principle of schools towards a meritocratic society, i.e. one based on ability and merit. This required that all children were given equal opportunities in getting an education and, consequently, employment. The legislation that really drove this belief came during the 1960s and 1970s with the growth of comprehensive schools. As Brown (1990) argues, this did produce a general increase in standards and was very influential in equalising gender inequalities. However, it still maintained other, relative, differentials in education performance. Several features of the second wave were perceived to have 'failed' to materialise. For example, standards in the lowest performing schools and areas still remained significantly low. The ideology of meritocracy was soon criticised for being simply 'symbolic', and it did not prevent the very high levels of youth unemployment that came with negative economic growth during the early 1980s.

Many advocates of change criticise the ineffectiveness of institutional powers to reconcile these problems. For example, Chubb and Moe (1988) discuss the effectiveness of the private sector in providing the desired quality education because they are controlled by society rather than by bureaucracy. Using Hirschman's (1970) ideas of 'exit' and 'voice' they also stress the greater social inequality that arises in a public system. This is because, they argue, the only way to change the education a child is receiving in the state sector is to use your 'voice' to bring about change within the school. Consequently this requires a particular level of cultural capital to be successful, and since cultural capital is unevenly distributed, inequalities in bringing about change will occur. Within a private education system parents can also use 'exit' by changing school completely in order to get the education they want for their children. This, Chubb and Moe (1988) stress, gives parents greater opportunities to change the education of their child and possibly requires less cultural capital to bring about. Having the 'exit' option is also seen as promoting greater responsiveness in schools in the first place so that parents are regularly considered when deciding what education to provide and how it is delivered.

From the perspective of public education, such advocates of change argue that the state schools are in a monopolistic situation. This can be, as Lauder (1991)[2] presents it, through the zoning system used to allocate pupils to schools promoted quite heavily during the second wave of reforms, or, as Tooley (1997) explains, not having any element of choice in the system promotes monopolistic behaviour by schools and the system generally.

These critiques of the second wave, and the changes to the political economy outlined above, led to the current, third wave, in legislation under the Conservative Government, 1979-97. Education legislation by the more recent Labour Government (1997 to present) has also continued this third wave of reforms.

The third wave in education legislation was organised around what Brown (1990) calls an ideology of parentocracy. In other words, the system is structured around the wealth and wishes of the parents rather than the ability and effort of pupils using the rhetoric of 'parental choice', 'educational standards' and the 'free market'.

The resulting policy shift in education has clear parallels with changes in the wider political economy. For example, the early changes to education policy by the Conservative Government focussed on resources, finance and the economy (David, 1989). These early stages to the reforms also introduced elements of 'privatisation' into the education system. Pring (1987) identified two major forms of this privatisation during the 1980s. The first was the purchasing at private expense of educational services that were typically paid for by the public sector. This is illustrated by the National Confederation of Parent Teacher Associations (NCPTA) who estimated that £40 million a year was being spent by parents on 'essentials' for schooling (Chitty, 1997). The second form of privatisation was the purchasing at public expense of educational services in private institutions. This dramatically increased with the introduction of the Assisted Places Scheme, which made bursaries available to more able children so that they could attend independent schools. In effect this provided subsidies to schools in the private sector of education that were otherwise financially struggling (Fitz *et al.*, 1986).

Within the Department for Education during the early 1980s, under the Ministerial leadership of Sir Keith Joseph, the most advanced form of privatisation in the education system was considered. The 'education voucher' has often been proposed by the New Right to provide 'a highly flexible instrument, with many variations, that would replace the financing of schools through taxes under political control and bureaucratic supervision by payments direct from parents thus equipped with a new ability (for the 95% with middle and lower incomes) to compare schools and move between them' (Seldon, 1986, p.1). However, Sir Keith was persuaded by civil servants, who studied the introduction of an education voucher, not to proceed with this policy because of its complexity (Chitty, 1997) leaving very little sign of any significant reforms being implemented.

The failure of the Government to introduce a decisive set of reforms in the first two terms of office was eventually ended with the Education Reform Act (ERA) in 1988. This Act signalled the beginning of the major part of education reform and heralded a significant shift in policy. The Department for Education outlined the objectives of the Act as:

> The Government's principal aims for schools are to improve standards of achievement for all pupils across the curriculum, to widen the choice available

for parents for the education of their children and to enable schools to respond effectively to what parents and the community require of them, thus securing the best possible return from the substantial investment of resources. (Department for Education, 1988)

In other words, the general aim of the Act was 'to introduce a more competitive, quasi-market approach to the allocation of resources in the education system, and to increase the range of parental choice over children's schooling' (Bartlett, 1993, p.125).

According to David (1989) the main thread running throughout the 1988 ERA was for greater financial accountability by devolving budgets and responsibility to schools as individual management units. This was supplemented with giving parents the opportunity to choose the school for their children to attend. A secondary thread tying together the many sections of this piece of legislation was the increasing centralisation of resource control and decision-making. These two threads were very important since subsequent legislation to the 1988 ERA has also been linked in these two ways. Whitty and Power (1997) have illustrated this by separating all the education reforms of the Conservative Government into these two categories of change; devolution and choice (Table 2.1), and centralisation (Table 2.2).

Table 2.1   Devolution and choice reforms, 1979-97

- Assisted Places Scheme, 1980 Education Act
- Reformed Governing Bodies, 1986 Education Act
- City Technology Colleges, 1988 Education Reform Act
- Grant-Maintained Schools, 1988 Education Reform Act
- Local Management of Schools, 1988 Education Reform Act
- Open Enrolment, 1980 Education Act and 1988 Education Reform Act
- Specialist Schools, 1993 Education Act
- New Grammar Schools, 1996 White Paper

Source: From Whitty and Power (1997, p.237)

Table 2.2   Centralisation reforms, 1979-97

- National Curriculum, 1988 Education Reform Act

- National Curriculum Council, 1988 Education Reform Act

- National Assessment, 1988 Education Reform Act

- School Examinations and Assessment Council, 1988 Education Reform Act

- Office for Standards in Education (OFSTED), 1992 Education (Schools) Act

- School Curriculum and Assessment Authority (SCAA), 1993 Education Act

- Funding Agency for Schools, 1993 Education Act

Source: From Whitty and Power (1997, p.237)

Over the period 1979-97 there have been four key areas of reforms that the Conservatives introduced to the state education system in order to develop devolution and choice. First has been the introduction of formula funding. Prior to this reform, administrators in the respective LEAs allocated budgets to schools, but under this new legislation a significant proportion of the schools' budgets were allocated according to the age and number of students that they had attracted. In 1990/91 56% of the overall General Schools Budget was determined by the number of pupils a school had. The second key area of reforms was the introduction of delegated budgets and delegated management under the legislation on Local Management of Schools (LMS). This allowed schools to decide for themselves how their budgets would be spent. Consequently, this gave greater powers to Headteachers and elected school Governors in deciding the shape and structure of education within their school.

These two areas of reforms, formula funding and LMS, have significantly altered the relationship between the LEA and the school. The powers and influence of the LEA have been dramatically diminished as a result of these reforms and in turn has given schools a great deal more individual freedom in providing education.[3]

Even though these first two areas of reform have reduced the role of the Local Authority it still maintains some influence on schools since, for example, it is the LEAs that set the total General Schools Budget. However, the third key area for the reforms gave schools the opportunity to completely 'opt out' of Local Authority control. If a school decided to 'opt out' then they would gain Grant Maintained (GM) status, which meant that their budgets would come directly from Central Government. Such schools could then operate entirely independently of LEA control, from their admission policies to the drawing up of contracts for their teachers. These 'new' schools should also be seen in relation to the introduction of City Technology Colleges, which were introduced to provide a different 'type' of education. Similarly to GM Schools, these Colleges were independent of LEA control, but they differed in that they were funded from both the public and the private sector, hence the closer link to employment, industry and technology.

The final area of reforms has enabled parents to make a preference for a school of their choice instead of being allocated a school by the Local Authority. Parents would then have the right for their children to attend their choice of school as long as the size of a school's intake did not exceed some predefined admission level. Such mechanisms were introduced in the 1980 Education Act although it was still possible for LEAs to ignore parents' preferences if it prevented them from 'making efficient use of resources'. Allowing LEAs to continue to set Planned Admissions Levels for each school based on their forecasting of future pupil numbers inevitably constrained 'parental choice'. However, the 1988 Education Reform Act loosened these constraints by imposing admission levels for schools called the Standard Number, which were determined by the size of each school's intake in either September 1979 or September 1988, whichever was highest. Coincidentally, the 1979 intake year was a peak year for secondary school admissions. Subsequently, there were much greater opportunities for parents to actually get a place for their children in a school of their choice. The establishment of an Appeals procedure giving parents the opportunity to have a hearing in front of an admissions panel, because they were dissatisfied with their final school allocation, supported this policy of open enrolment. Another feature of the reforms that also supplemented the opportunity for parents to choose schools, and that has received widespread publicity, particularly in the media, has been the publication of examination performance league tables. These were intended to provide useful information to parents to assist them when they were choosing a school for their children.

Together, these four key areas of reform have created a system in which parents are encouraged to behave as 'consumers' of education and where schools are given the opportunity to respond to the demands of these 'consumers' in order to maximise the number of pupils attending their school and, therefore, maximise the level of resources they receive. This amounts to the two key concepts of the market, that of choice and competition, but also where the Government is still the purchaser. Consequently, it is difficult to deny that these education reforms have produced a quasi-market in education. As Lauder (1991) argues, the creation of a market in education has moved education from the political sphere into the economic sphere.

Similar education reforms have been introduced in other countries typically because of the same underlying pressures in the political economy (see for example Ball, 1998) and because of policy networks (Whitty and Edwards, 1998). Examples can be found in The Netherlands (Dronkers, 1995), New Zealand (Waslander and Thrupp, 1995), the USA (Chubb and Moe, 1990), Scotland (Adler *et al.*, 1989), France (Ambler, 1994), and Israel (Menahem *et al.*, 1993). However, reforms to the English and Welsh education system stand out as being more advanced towards the New Right's 'ideal' scenario for educational provision (Chubb and Moe 1992).

It is also important to note that these reforms did not end when Margaret Thatcher stood down as Prime Minister and leader of the Conservative Party in November 1990. The subsequent Government under the leadership of John Major produced a White Paper 'Choice and Diversity: a New Framework for Schools' (1992). This established mechanisms to create new types of schools at the secondary level intended to provide greater diversity in the school system. Then the 1993 Education Act provided the legislation for the establishment of specialist schools, which allowed maintained schools to establish themselves as 'specialists' in teaching particular subjects, such as technology and languages. Such reform to extend the 'choice' of schools for parents was continued in 1996 with another White Paper proposing the (re)establishment of grammar schools. The Conservatives fought the 1997 General Election with a manifesto promising that there would be a 'grammar school in every town'. However, grammar schools were one of the losers when the Labour Party won the 1997 General Election, as they rejected plans to create 'new' grammar schools and instead gave parents the opportunity to vote to remove grammar school status from local schools.

The first Education Bill the Labour Government introduced, only a few months after being elected, ended the Assisted Places Scheme which helped them to increase spending in education by an extra £19 billion over three years. However, apart from these specific changes to the Conservative legacy, the Labour Party has embraced 'public sector managerialism' (Cutler and Waine, 1997) by using a similar language to the Conservatives of 'targets', 'monitoring' and 'performance'. The 'Excellence in Schools' White Paper (July 1997), and the subsequent 'School Standards and Framework' (SSF) Act (1998), included the setting of further standards and performance targets for schools, Local Authorities and Central Government. It also proposed to modernise the comprehensive principle by continuing to offer school diversity but without returning to the 11+. The SSF Act also introduced Fair Funding, which, again, attempted to improve the allocation of resources to schools based on their success in the education market place.[4] The most radical of the Labour Government's plans did not appear until a Green Paper in December 1998 which proposed performance related pay for teachers – extending the market principle by providing incentives for success.

The discussion now turns to summarising the literature for an evaluation of these reforms in education before proposing a rationale for this study.

Evaluation of Education Reforms

Before specifically focussing on an evaluation of education reforms it is useful to examine the literature on critiques of changes in policy for public services generally. As Pierson (1994) concludes, the 'retrenchment' policies of both the Reagan and Thatcher programmes fell short of their original aspirations. For example, while the UK Government managed to reduce spending as a proportion of GDP between 1979 and 1989 they were unable to prevent spending from rising from 1990 onwards. Consequently, by 1996, spending levels were the same as they were in 1979 (Taylor-Gooby, 1998). Another significant feature of the introduction of the neo-liberal ideology to the Welfare State has been the extent to which State powers have been reduced. As Clarke (1996) argues, the institutional role of the Government has only been changed and not reduced, which is reinforced by society's continued support for core public services (Taylor-Gooby, 1998). Clarke (1996) identifies three ways in which state power is now exercised:

i.   Greater centralisation of financial control.
ii.   Objective-setting and evaluation.
iii. Dispersal of 'agency' through sub-contracting and delegated authority.

However, it could be argued that such powers were intended to remain with the State if the reforms were to resemble that quasi-market approach to providing public services. Therefore, an evaluation of the reforms must be considered in the context of the quasi-market model. At a theoretical level Bartlett and Le Grand (1993) provided four criteria, or objectives, for evaluating quasi-markets:

i.   Efficiency
ii. Responsiveness
iii. Choice
iv. Equity[5]

In terms of the 'market structure' of a quasi-market it was deemed that competition is necessary for both purchasers and providers. The exceptions to this would be where either of these is in a monopoly situation that could not be broken. Secondly, 'information' is necessary to give providers and purchasers access to 'accurate, independent information, providers primarily about costs, and purchasers about quality' (Bartlett and Le Grand, 1993, p.33). The third condition is to provide some form of 'motivation' for providers and purchasers. The problem here is to generate some element of motivation for the provider when they are non-profit organisations. The fourth condition is to prevent 'cream-skimming', in other words, to prevent incentives from arising for purchasers and providers to discriminate between users. The final condition is to maintain and possibly improve the 'quality' of the product the quasi-market is delivering.

On this basis quasi-market reforms in some areas have been more successful than in others. For example, Le Grand and Bartlett (1993) suggest that reforms in education and housing have been more successful than the introduction of quasi-markets in health care and social care. However, at a broader level the concept of the quasi-market has been criticised for providing too many contradictions in policy. Cutler and Waine (1997, pp.85-86) argue that quasi-markets generate three sets of contradictory pressures:

Consumer choice   vs.   Budgetary constraints and efficiency
Decentralisation and diversity   vs.   Regulation
Maintenance of quality   vs.   'Value for money' and efficiency

Hudson (1992), who evaluated the quasi-market in British health and social care, sees the contradictions that are generated as providing a conceptual problem in which the relationship, between the quasi-market and the necessary planning and regulation of a public service, needs to be reconciled. These structural problems of the quasi-market in general are also compounded by a more behavioural critique of this model of provision. As Taylor-Gooby (1998) explains, the market model assumes that consumer self-interest will permeate choice in the market, but evidence suggests that social and cultural values play a very important role in influencing the behaviour of consumers of public services.

It has always been acknowledged that there are advantages and disadvantages of privatisation in public education (Mulford, 1996), but it is necessary to examine the specifics of the quasi-market in education to make a useful evaluation of the reforms.

A preliminary evaluation of the education quasi-market by Bartlett (1993) using the five conditions, as outlined by Bartlett and Le Grand (1993) above, was promising. In terms of the 'market structure' Bartlett suggested that 'the market structure is broadly, if imperfectly, competitive, and that any failure of the quasi-market system in education is unlikely to be associated with an absence of alternative providers' (1993, p.139). However, there is no contingency in the reforms for entry to and exit from the education market, i.e. there is no provision for the closing and opening of schools. It is also difficult for schools to expand in size because of two pressures preventing this; lack of money for capital expansion and a reluctance for having large schools (Levacic, 1994).

'Information' in the education quasi-market is very important since school choices are based on judgement and not actual experiences. As Bartlett warned 'the quasi-market mechanism will not work well where imperfectly informed consumers based their choice of school on non-educational characteristics such as the predominant social class or racial composition of the school intake' (1993, p.144). Publication of school prospectuses and examination performance league tables has assisted with providing information but there remains great debate about what information should go into the examination performance tables.

In terms of 'motivation' Bartlett (1993) believes that both the LEA (the purchaser) and the individual schools (the providers) have retained the importance of quality education and have used this to drive their decisions in the market. The fourth condition, of preventing 'cream-skimming', is more problematic. As Bartlett (1993) explains, the cost of educating children varies according to their individual needs, therefore schools are more likely to prefer pupils who are likely to produce the greater returns, i.e. examination performances, on the investment by the school. Such pupils are more likely to be from socially advantaged backgrounds. This has important ramifications for popular schools because they can begin to be selective over their admissions, 'and by encouraging an increasingly selective admissions policy in such schools, open enrolment may be having the effect of bringing about increased opportunity for cream-skimming and hence increased inequality' (Bartlett, 1993, p.150). As Levacic (1994) notes, this changes the 'motivation' for popular schools from just getting a particular number of pupils into their school to getting a particular type of pupil into their school. The issue of 'cream-skimming' dominates the literature on education reforms, and as Whitty (1997) stresses, most evidence suggests that it exists. Analysis by Bush *et al.*, (1993) suggested that 30% of comprehensive schools studied used covert means to select their intake. On the other hand, Levacic (1994) provided evidence to indicate that there were no 'cream-skimming' mechanisms in operation other than for religious, or voluntary aided, schools, which have always had some form of selection in their admissions policies. However, the academic literature tends to lean towards the presence of 'cream-skimming', in theory if not yet in reality. As Ranson (1988) suggests, the market 'pretends' to be neutral and in fact confirms and reinforces the pre-existing social order of wealth and privilege. The extensive work of Stephen Ball has begun to show that the process of choice in the education market is socially divisive (see for example Ball *et al.*, 1995) and, most importantly, that the purpose of a quasi-market in education is to provide a class strategy for middle class reproduction (Ball, 1993, p.1998).

The last of the five conditions, the question of quality, has been neglected in the literature, simply because the reforms have not been in place for long enough to examine the relationship between the education quasi-market and changing education standards. But Levacic (1994) indicates two ways in which these reforms, particularly LMS, can impact upon the school classroom. The first is by finding more cost-effective ways to deploy a given quantity of resources, this in turn, would maximise the school's own investment. The second way would be to target and assist those elements that improve school effectiveness and school improvement. It has certainly been shown that Headteachers and governors have generally welcomed their new devolved responsibilities and budgets, and list the benefits in the following order (Marren and Levacic, 1994):

i.   Greater school autonomy.
ii.   Improved financial awareness in schools.
iii. More efficient use of resources.
iv.   Greater flexibility in resource use.
v.   Improved planning within schools.

However, school autonomy has been severely restrained by the implementation of the National Curriculum (see Table 2.2 above), which reflects the neo-Conservative tendencies of the Conservative Government during the 1980s and 1990s. The National Curriculum prescribes the type of education a school must provide, therefore preventing schools from changing the product they are offering to the consumers. This contradicts part of the reforms that tries to offer diversity and choice in the education system (Cutler and Waine, 1997). A study of grant maintained schools, introduced to provide diversity in the market place, by Fitz *et al.* (1993), also found that there was little change in the curriculum and pedagogy of these schools from other LEA-maintained schools.

Similar to quasi-markets in general, the education market has been criticised for generating another contradictory pressure in policy between choice and efficiency/effectiveness (Cutler and Waine, 1997). This potential contradiction can be seen in the extent to which efficiency objectives are passed over in the pursuit of giving parents choice in the education market. This is because choice is heavily constrained by the availability of places within popular schools. Greater surplus tends to allow greater choice, whereas a more efficient education system, with relatively small surplus places, inhibits potential choices of schools. Consequently, to provide a more efficient education system there would be a need for greater planning which inevitably would have to prescribe admission arrangements. On the other hand, as Thomas (1996) believes, it could be argued that education can never be economically efficient because the production possibilities that can control efficiency are limited.

To conclude this evaluation, it becomes very clear that there are two key concerns of the education quasi-market. The first is that there are many contradictions in the policy strategy, say, between choice and efficiency, or between centralisation and devolution. The second, and probably the most important in terms of the role of education in society, is that the quasi-market appears to be producing inequalities between social classes, because of the process of choice, and between schools, because of the allocation of resources. Proponents of reform generally do not shy away from these conclusions. Instead they tend to argue that the reforms have not gone far enough. For example, Tooley (1997) proposed three further reforms that would reduce inequity in the market place:

i.   Current allocation of resources to the disadvantaged should be made more obvious.
ii.   League tables should be abolished and some form of consumer report be used.
iii.   Measures were needed to identify failing 'sink' schools early and remedies should be enforced.

Similarly, Moe (1994) focussed on the inability of the system in being able to change the supply of schools controlled by a LEA and proposed that the solution was to give more autonomy to schools to allow them to expand or contract at will.

Just as these reforms are only the beginning of the 'third wave' in education legislation, research studying and critiquing the reforms is also only beginning to understand the processes and patterns of the 'new' education market. It is here where this study begins. This Chapter concludes by discussing the requirements of research on education quasi-markets generally, and how this study attempts to meet some of those requirements.

Rationale For This Study

The main aim of this Chapter has been to outline the status of research into the new education market, but more detailed analysis of the research is presented alongside this research in forthcoming Chapters. The purpose of this final section is to provide a rationale, based on the literature, for the approach and objectives chosen for this study.

Research examining the processes and patterns of the 'new' education market has been relatively healthy within the academic fields of sociology and education. However, there has been little 'geographical' analysis of this new form of education provision. One of the key aims of this study is, therefore, to bridge the geography-education disciplines so as to achieve two objectives; first, to provide a geographical perspective on an issue of great importance within education, and secondly, to bring the debate on education policy into the realms of geography.

In terms of the main objectives of this study it is useful to consider Ball's (1997) template of research positionality for policy science in general. Ball suggested that in terms of education policy research the most important tension is between efficiency concerns and social justice concerns. As Glatter *et al.* (1997) also conclude, the most important implications of the education reforms during the 1980s and the 1990s are on school admissions and social equity.

The central feature of this research, therefore, is to examine the education quasi-market from a perspective associated with concerns for social justice. It will focus on inequalities and inequities[6] in the 'new' education market, and will examine these inequalities at four different spatial scales and levels of the quasi-market: the national level of education provision, the quasi-market at the LEA level and the School level, and the market process at the level of the individual household. The spatial framework of the education quasi-market and its four interrelated levels of activity is presented and discussed further in the next Chapter. Providing a comparative approach to the research, in which different 'contexts' are considered to see how they influence the process of the market differently, further enhances this. For example, this study sets out to compare different LEAs, different schools and families from different social backgrounds. It also attempts to compare different geographical locales, for example, urban and rural, to see how spatial constraints might influence the effectiveness of the education reforms.

Le Grand (1991) discusses the economic conceptions of equality and equity and argues that they tend to be considered as the 'end-result' of a process. For example, these economic conceptions would propose that education quasi-markets produce inequalities in the final allocation of pupils to schools, e.g. social segregation of intakes. But, Le Grand (1991) concludes that it is important to focus on inequalities **in** the process, i.e. inequalities in the process of choosing a school. 'A society with less inequality in choice sets will be one with less inequity; the challenge for policy is to move in the direction of greater equality of choice, and hence greater equity, without too serious a compromise of other values' (Le Grand, 1991, p.101). However, Halsey *et al.* take a different perspective, 'much of the case against educational markets rests on studies which observe social-class and ethnic inequalities of choice. However, the mere observation of inequalities of choice is not a persuasive argument against markets unless it can be shown that these create greater polarisation than occurred under a zoned system' (1997, p.359). Consequently, since this study wishes to investigate inequalities and inequities in the new education market it will focus on both the process-side and the product-side of the market.

When examining the process-side of the education market it is necessary to try to understand the behaviour and decisions of the families concerned. Taylor-Gooby (1998) proposes three areas of development when studying economic behaviour generally, and each relates to potential inequalities in choice:

i.   Human limitations of choice.
ii.   Rationality and consistent preferences.
iii. The social context of choice.

Such a behavioural approach is very applicable to the study of the education reforms and the quasi-market, and in particular the way parents and their children choose a school within a new system of school provision. This is an area of research that has seen a great deal of activity, for example, the work of Ball and his team has been very influential in understanding the process of choice. However, this research does not compare the process of choice in different contexts, as Bowe *et al.* admit, 'it remains our view that the context of parental choice-making remains an underdeveloped area of social enquiry' (1994, pp.38-39). An example of why it is necessary to examine the education quasi-market in context is related to the number of surplus places within a LEA. As Dennison (1983) and Thomas and Robson (1984) have shown, school rolls are constantly changing due to the underlying demographic changes in the population of a particular locale. The resulting changes in the number of surplus places in schools will have an effect upon the nature of competition between schools and the allocation of pupils to schools. Consequently, one of the main arguments for a geographical perspective using different spatial scales of the education market is to provide the necessary context to choice at the household level. In terms of the Ball's (1997) dichotomy of research positions it would be fair to say that this study is 'context rich', 'multi-levelled' and with a 'linked focus'.

One particular aspect of a 'context rich' approach that permeates through this study is the various constraints that affect the performance of the quasi-market. Other research tends to focus simply upon the social constraints on choice, such as the varying levels of cultural capital that families have deemed necessary to engage with the market (see for example Gewirtz *et al.*, 1994). However, spatial or geographical constraints are often neglected. This is probably because of two reasons; most research tends to be based on London Boroughs where access to a number of schools is not heavily constrained by access and distance, and simply because few geographers study the education market.

There has been significantly less research examining the product-side inequalities of the 'new' education market, generally because larger and more empirical data sets, such as admissions for **all** schools in a Local Authority, are required. It is therefore not surprising that there has been a call for such research in the literature. For example, in a recent review of the quasi-market development (Bartlett *et al.*, 1994) it was argued that one of the key areas for future quasi-market research was that they be empirically based. Similarly, Glatter *et al.* (1997) specifically suggest that one of the aims for future research on the education market was to monitor school intake patterns and to provide empirical evidence for the effects of the market on social composition of intakes. This study has access to such data sets and can therefore provide a useful analysis of inequalities as a result of the market process.

This research takes a juxtaposition between the contrasting research perspectives indicated. So, for example, this study is generally 'practice oriented', i.e. based on outcomes of the policy reforms, and, as indicated above, it is 'context rich' and focussed on 'social justice'. This study also hopes to provide a 'conceptual' approach to the study of the education quasi-markets in order to overcome the complexities of choice and competition. It also intends to provide a 'critical' analysis, both with policy and the academic debates. One position of this research that does limit its understanding is that it is 'atemporal', or rather, it is just a snap-shot of a dynamic and changing process. Other positions are rather more complex to define, largely because this study examines four spatial scales of the market process. Consequently, it studies the process at each 'single level' but also provides 'multi-level' analysis. It takes a 'national/local' position but also takes the opportunity to provide 'general' conclusions from the findings. Finally, in terms of developing policy the study concludes with 'policy implications' and suggests further developments to the reforms. However, whether they are heard will be outside the control of this research.

Notes

1   Changes in the provision of education throughout the 20th Century are discussed in more depth in Chapter 3.
2   Lauder (1991) does follow this by criticising the market system for not providing a democratic education system.
3   It has been argued that the role of the LEA is still significant. This is discussed in more detail in Chapter 3.
4   The SSF Act also established new guidelines for admissions that will constrain the market in future. But since these were introduced after this research had been undertaken the significance of these proposals is discussed towards the end of the book in Chapter 8.
5   Bartlett and Le Grand define an 'equitable' quasi-market 'to be one where use is determined primarily by need and not by irrelevant factors such as income, socio-economic status, gender or ethnic origins' (1993, p.19).
6   The difference between equity and equality used here is based on Le Grand's (1991) discussion - equality has a descriptive component, whereas equity is a purely normative concept (see also Smith, 1994; Hay, 1995).

3 The Geography of Education and the Education Market

Introduction

In order to consider the processes at work in the 'new' education market place it is important to examine the context that underlies them. According to Ball (1986) there are two underlying facets to the changing structure of educational provision – the political nature of change and the relationships with social class divisions. However, rather surprisingly, Ball overlooked another key feature of the education system in England and Wales – many of the education market processes operate at different spatial scales. One of the most significant agents in the education market is the LEA, which generally operates at the regional scale. This Chapter will, therefore, focus on the education market place at the regional or administrative level in order to provide a context to the more detailed processes at work in later Chapters.

The discussion in this Chapter represents a more 'traditional' geographic approach to education exemplified by Burdett who defined the geography of education as 'the variation across the country in the types of institutions, the levels of funding, the nature of access, the curriculum offered and the educational outcomes achieved' (1988, p.208). Even though this might be a rather simplistic view of the geography of education it serves as a very useful way of considering the state of education provision across England and Wales. But, of course, as Bradford (1990) highlighted, it is also necessary to understand the processes that affect the changing geography of education.

This milieu of education has been studied in some detail (see for example Bondi and Matthews, 1988), but this analysis will focus on the ways in which the regional balance of provision and policy influences the education market. As mentioned earlier the LEA plays a very important role in the education market both as a provider and as an arranger. This bureaucratic role is embedded into the education geography through their particular approach to policy and consequently defines the sphere in which most of the market choice and competition in education occurs. This does not mean to say that this 'market place' is totally bounded and finite, for example, in 1992 in England and Wales 186,000 students crossed LEA borders to attend school. More recently this has been tested in the law courts upon which the Greenwich Judgement (1989) established that maintained schools could not give priority to students who live in the same LEA. As a result the School Standards and Framework Act (1998) issued guidelines for Local Authorities to work closely with neighbouring authorities in school admission arrangements. Ball *et al.* (1994) have also suggested that the geographical boundaries that define the LEA are totally arbitrary and that the areas of choice and competition are defined differently. However, for the vast majority of parents and students the LEA is all encompassing when it comes to choice. Table 3.1 shows, for a sample of LEAs, the proportion of students that lived outside the Local Authority boundaries.

For the Inner London borough (Eastern) over 18% of the intake came from outside the Local Authority, similar to the intake in the other Inner London borough (Western). The Outer London borough (Northern) also had a high proportion of its intake from other LEAs, but very few of the other LEAs, urban metropolitan or shire counties had such a significant number of students crossing LEA boundaries to attend a school of their choice. Obviously these examples did not account for the number of students who lived in these LEAs and decide to choose a school in another Local Authority. It also did not take into account the distribution of the population around the boundaries, but just by considering the relative size of the perimeters for each Local Authority only the Inner London borough (Eastern) really stood out. So, with the possible exception of the London boroughs, the LEA as a geographical unit did act as a bounded market place for choice and competition.

Since it has been argued that the spatial extent of the Local Authority provided the arena for the majority of choice and competition, i.e. the market place, then this geographical space provides a useful unit for analysing the regional mosaic of education, and, consequently the backdrop to the education market. Apart from the majority of education data being aggregated at the LEA scale there are also three main reasons as to why the LEA is an important feature of the education market. Firstly, since the ways LEAs operate and organise their schooling differ from each other then the many market places that exist are never homogenous. Therefore, the role and actions of the LEA are very important to understanding how the market might function and are consequently discussed in more depth in the section on the 'institution space'. The second feature of LEAs, of significance here, is how schools themselves differ from each other within their Authorities.

Table 3.1   Cross-LEA choice for selected LEAs

LEA	Proportion of total intake (%)	Adjusted figures*
Inner London borough (Eastern)	18.17	8.62
Inner London borough (Western)	18.82	3.81
Outer London borough (North)	13.35	2.84
Large metropolitan borough (West Midlands)	2.89	3.31
Small metropolitan borough (West Midlands)	4.84	2.69
Metropolitan borough (Greater Manchester)	4.67	1.80
County (Eastern)	2.12	0.35
County (West Midlands)	4.28	0.48

*   The proportions are recalculated to reflect the different size of perimeters for each LEA. In the case of the Eastern county, only the land locked perimeter is used.

This 'diversity of schooling' is discussed in two sections, the first examines the different types of schools that exist and the second considers the geographical distribution of this school diversity by LEA. The final aspect in outlining the context to studying the education market is the marked variation in parental empowerment seen in the differing rates of appeals lodged by parents when they do not get their choice of school.

This Chapter considers each of these features in which the education landscape differs and how that might impact on the final level of activity in the market place, and the way the market may operate and function. But first it is necessary to consider more general variations in education provision and outcomes across England.

Spatial Variations in Education Provision and Outcomes – Some General Comments

Burdett (1988) has claimed that the 1988 Education Act could well exacerbate spatial variations in provision and in turn would result in a spatially uneven education market. In other words, the spatial variation in provision will influence market processes and that these same processes will in turn change or reinforce the spatial variation in provision. Consequently it is pertinent to begin by presenting a general picture of the geography of education using a number of different features of this landscape before moving on to examine the features that will have a direct bearing upon the market place as outlined above.

The first important feature of this education landscape is the distribution of expenditure by Local Authority. Education funding has changed significantly in the last decade. For example, the distribution of the standard spending assessment (SSA) per pupil for 1996 shows an obvious urban bias to funding alongside a relatively high level of funding for LEAs in the South East (Figure 3.1). On the other hand the introduction of formula funding has begun to spatially level out the base expenditure for education (Figure 3.2), but the South East still has a relatively higher level of funding per pupil than other LEAs.

This has helped to illustrate the spatial variation in resource distribution, or 'inputs', but at the other end of the education process there are also spatial variations in the resulting 'outputs'. For example, the proportion of pupils aged 15 who obtained 5 or more GCSEs with grades between A and C in the 1995-96 academic year was above average in the South contrasting with the generally low level of achievements in urban authorities (Figure 3.3).

These spatial variations can be confirmed by non-parametric correlation coefficients between variables (Table 3.2). So for example, as the density of schools in an authority gets larger then so does the level of SSA expenditure. This funding is also related to the levels of deprivation calculated by the DfEE (Figure 3.4). The more recent equalisation of education expenditure can be seen with the smaller correlation with school density. However, as a result, the relationship of expenditure to levels of deprivation has also fallen. It should be noted that this is meant to be offset by new additional budgets and funding formulae to meet the varying needs generated by levels of deprivation that are calculated on top of the base levels of expenditure.

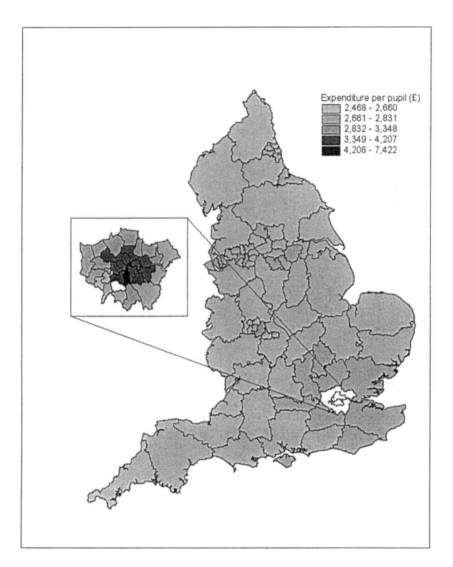

Figure 3.1 Standard spending assessment (SSA) per secondary school pupil, England 1996-97

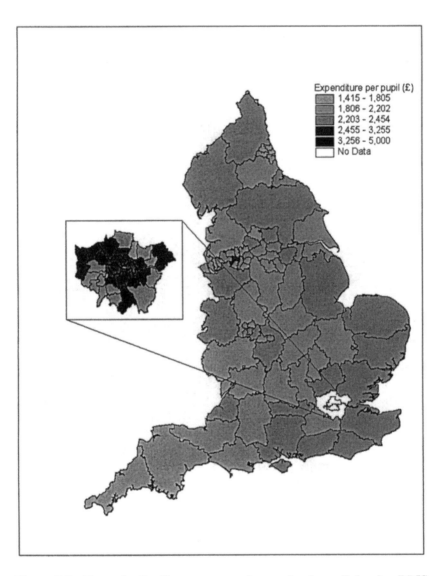

**Figure 3.2   Formula funding per secondary school pupil (under LMS
formula), England 1998-99**

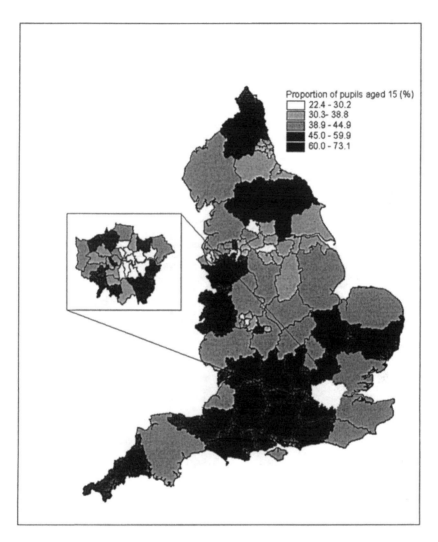

Figure 3.3   Proportion of pupils aged 15 obtaining 5 or more GCSEs with grades A to C, England 1995-96

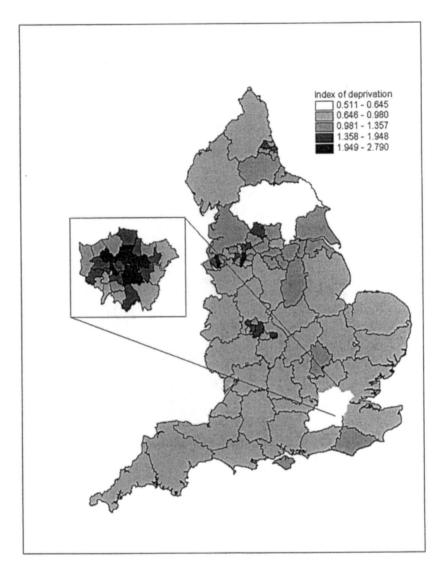

**Figure 3.4   Additional educational needs (AEN) – index of deprivation
used by the DfEE, England 1998-99**

These levels of funding can impact upon the quality of education since there is a relationship between levels of funding and, for example, pupil-teacher ratios (R = -0.715) The resulting variations in quality of education can then be seen to relate to the outcomes of the education process. But ultimately, any allocation of resources should be distributed to alleviate the obvious relationship between levels of deprivation and the GCSE examination performance of pupils in a LEA (R = -0.774). Areas with relatively high levels of deprivation appear to also produce the relatively lower levels of GCSE achievement even though there has been some targeting of resources to these areas.

One other feature of the education landscape that will have an impact upon the education market place is the varying number of surplus places to be found within each Local Authority. On the one hand, the market in education was introduced to make the education system more efficient by reducing the levels of surplus places in Local Authorities (Audit Commission, 1996). But, on the other hand, the surplus allows there to be greater movement of pupils between schools without schools becoming too over-subscribed. According to the proportion of surplus places in secondary schools for 1988 and the proportion of schools in each Local Authority that have more than 25% of their total capacity being surplus, also for 1998, there is no spatial pattern across England (Figures 3.5 and 3.6). There appeared to be no relationship between the levels of surplus places in schools and the geographical nature of the authorities, be they north or south, metropolitan or county (see also Audit Commission, 1996).

Table 3.2   Change in the relationship between the density of schools and levels of deprivation with allocation of funds for education across LEAs

School Density and Deprivation	Funding mechanism	Non-parametric correlation co-efficient (R)*
Density of schools	SSA Expenditure	0.693
	Formula Funding	0.542
AEN Index score (DfEE deprivation measure)	SSA Expenditure	0.707
	Formula Funding	0.421

* Correlation co-efficient (R) based on data from all English LEAs.

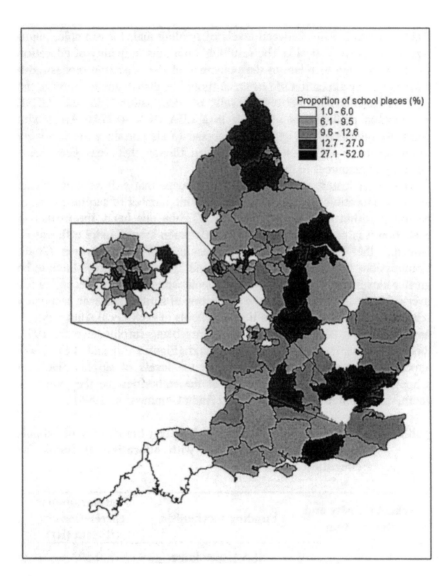

Figure 3.5   Surplus places as a proportion of total secondary school places, England January 1998

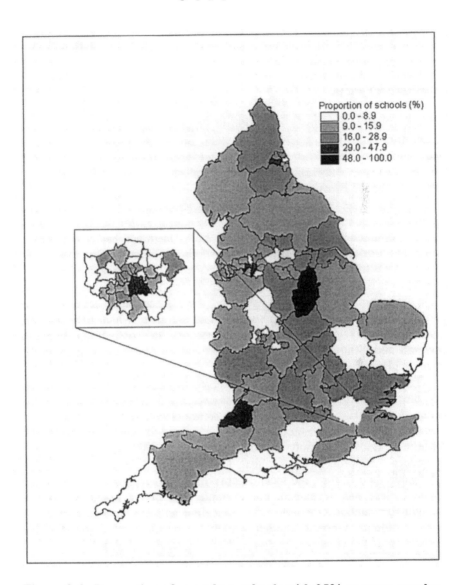

Figure 3.6   Proportion of secondary schools with 25% or more surplus places in them, England January 1998

The discussion now moves on to focus on features of the education landscape that directly influence the processes at work in the market place.

Institution Spaces

The first key step to mass education in England and Wales was the 1870 Education Act. This established School Boards to provide elementary education for those children who were not being provided with education from other voluntary and philanthropic means. In 1902 LEAs replaced these School Boards. The new administrative organisations were empowered to set up their own secondary schools and ever since have been given many powers to provide and run the education for the vast majority of the population. The power that these LEAs have had has changed over time and arguably it was at its peak towards the latter half of the 1970s. Irrespective of their current roles, the decisions and approaches taken by LEAs over time has shaped and moulded the market places that now exist. The variations in the decisions made and the different approaches taken to policy, alongside the differing socio-spatial arenas in which they operate, have inevitably created different approaches to competition between schools and different levels of encouragement for parents to engage in the market place.

The relatively small amount of research specifically focussing on differences in LEA practice has emphasised the differences created in the education market. For example, Ball *et al.* (1994, p.92) compared two LEAs and found that 'the market structure and dynamics in these two localities remain strongly framed and guided by LEA policies and practices'. Ball *et al.* argue that LEAs act as 'corporations', which alongside their history and local complexities define the way in which the market place has developed. For example, one of the Local Authorities sought to maintain comprehensive education and the class advantage of local residents, whereas another sought to 'deconstruct' comprehensive education and to create a new diversity among schools (Table 3.3).

Table 3.3   Summary of LEA variations in approach to the education market

Local Education Authority A	Local Education Authority B
Social Democratic	Conservative
Pro-comprehensive	Pro-selection and specialisation
Cartel of schools	School competition
LEA schools	Mixed (LEA/GMS/CTC)
School co-operation	School suspicion
Inflow problems	Outflow problems
Link primary scheme	Open enrolment/choice
Limited diversity	Planned diversity

Source: From Ball *et al.* (1994, p.92)

Brown (1996) simply compared the education system of England and Scotland, but the findings from this research can be used to highlight variations that inevitably exist between Local Authorities within England. It was suggested that Scottish regional Authorities are generally more powerful than their English counterparts and consequently while there were 633 grant maintained (GM) schools in England in 1995 there was only one in Scotland. Other than the levels of power between these LEAs, Brown also suggested there were three other features that distinguished the Scottish experience from the English experience of the education market at a regional level:

i.   Rural isolation.
ii.   Commitment of parents and professionals to education at a local comprehensive school.
iii. Lower level of Independent education.

All of these features can also be seen to distinguish between English LEAs. Some are shire counties and some are urban metropolitan LEAs. Some have had a stronger commitment to comprehensive education (e.g. Leicestershire County Council) than others. And there is just as great a variation in the provision of fee-paying schooling between English LEAs as there is across the English-Scottish border.

These examples have shown that there are variations in the approaches taken by LEAs to the education market. This discussion will now focus on two key elements of the political unit of the LEA that have a direct relationship with the way the education market might operate. The first is the spatial construction of the market place and the second is the immediate power and authority that the LEA has on the market place.

LEAs as Geographical Spaces

The first element that can be used to define variations between Local Authorities is the spatial nature of the LEAs. There are great differences in the sizes of the 109, pre-local government reorganisation, English LEAs. However, the spatial nature of each market place is intrinsically determined by the spatial distribution of schools and the population. As Kirby (1982) notes, secondary schools are organised to meet two constraints; first, they have to be relatively large enough so that certain levels of provision such as the curriculum can be offered or maintained and, second, they cannot be too large since they need to be evenly located across space, relative to the population.

In the Audit Commission's (1996) national report on allocating school places there were five different categories of LEAs identified, based entirely upon their geographical nature:

i.   London boroughs
ii.   Metropolitan districts
iii.   Urban counties
iv.   Semi-rural counties
v.   Rural counties

However, this classification still does not truly reflect the nature of the distribution of schools within the Local Authorities. For example, there is no such thing as a totally 'rural' LEA. All Local Authorities have urban areas within them. But even these do not reflect whether the schools are clustered around the urban area or evenly distributed across the urban and rural parts of the authority. The differing geographical distribution of schools and students will affect the territorial justice in accessing schools.

The size of each school's catchment area provides a more accurate reflection of the spatial unevenness of school distributions. The most convenient way of calculating these was to generate thiessen polygons[1] around each school within the Local Authority. Averaging the spatial extent of these catchments then provided an aggregate indication of the distribution of schools in the Local Authorities (Figure 3.7).

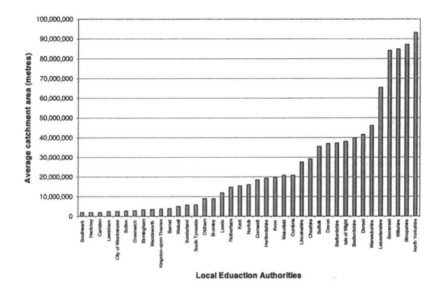

Figure 3.7 Average size of school catchment areas* for a sample of LEAs

* Based on Thiessen Polygons generated in ArcInfo.

Ultimately these spatial variations both between and within LEAs will control and influence the choice available to parents and students. However, as Kirby (1982, p.47) states 'spatial organisation is only meaningful when related to the social process that manipulate it'. Therefore, the next section examines the LEAs as seats of authority in which key decisions are made.

LEAs as Authority

Education legislation throughout the 1980s and into the early years of the 1990s has shifted power from the LEAs to the schools themselves, through local management and formula funding. However, as Mann (1989) explains, LEAs still play an important role. First, the LEA determines the funding formula in all cases except where the proportion of GM schools in the authority exceeds 10% of all schools. Once the number of GM schools reached this threshold then the running and funding of the schools in the authority was shared with the Funding Agency for Schools (FAS). However, the subsequent Labour Government abolished the FAS, along with the removal of GM status for schools. Many GM schools are now foundation schools and, significantly, have kept some control on their admissions.

The second role of LEAs, that Mann (1989) believes retains power, is the services that the LEA offers. Even though schools now place these services out to tender many LEAs are still the longest serving and biggest providers of these services. Another role the LEAs have is the statutory responsibility for delivering the National Curriculum, which they do alongside the headteachers and governors of their schools but they are still the dominant agency. Mann (1989) goes on to say that even GM schools were linked to the LEA since their budget was based on the funding formula that the LEA has for its own maintained schools.

As suggested earlier Williamson and Byrne (1979) have shown that in the late 1960s and early 1970s there were considerable spatial imbalances in educational resources and levels of education provision. These have then been linked to the varying levels of attainment throughout the country (Byrne *et al.*, 1975). But of all the current policies undertaken by the LEA their single most important role in the education market place is control of schools' admissions.

There are many different approaches to the admissions procedures (Mayet, 1997) that range from using fairly strict catchment areas as initial indicators of which schools children should attend, to having completely open systems of school admissions in which parents can list their preferences irrespective of any geographical boundary. Morris (1993) in a survey of all LEAs found that the distribution of Local Authorities between these two extreme forms of admission procedures was dependent upon the geographical characteristics of the authority (Table 3.4). It is clear that the London Boroughs were more likely to operate open preference systems whereas the more rural County LEAs were still using catchment systems for their admissions.

However, admission procedures are constantly changing. Forrest (1996) found that LEAs were abandoning catchment systems in favour of open preference systems. In 1985 61% of LEAs operated secondary catchment systems, but by 1996 this had fallen to 41%. Forrest (1996) also confirmed the findings of Morris (1993) that County LEAs were more likely to be operating a catchment model of admissions.

Table 3.4   Proportion of schools by admission arrangements and LEA type*

	Catchment arrangements		Open arrangements	
	Primary (%)	*Secondary (%)*	*Primary (%)*	*Secondary (%)*
County LEAs	75.8	74.3	24.2	25.7
Metropolitan district LEAs	45.5	63.6	54.5	36.7
London borough LEAs	25.0	17.6	75.0	82.4
Total	54.9	58.1	45.1	41.9

* Data from a survey of all LEAs. Breakdown of respondents as follows: County LEAs = 33; Metropolitan district LEAs = 22; London borough LEAs = 16; Total LEAs = 71 (109 English LEAs in total).

Source: Based on Morris (1993)

Interestingly Jowett (1995) identified four different models of admission procedures that go beyond the simple distinction between catchment and open systems: selective secondary schooling admissions; where parents make direct applications to schools; where parents state their preference(s); and where parents either accept allocated school or make an request for an alternative school. Mayet (1997) also acknowledged the importance of geography on the admissions process.

Whichever system of admissions is used within an authority there is a direct relationship with the way the market will operate. This can be through either the ease with which parents can opt to choose a school other than the school they would have traditionally been allocated (i.e. the level of cultural capital required for choice) or the level of dissatisfaction that arises from not getting the first choice of school. For example, Forrest (1996) makes the link between the type of admissions systems in operation and the number of Appeals over school allocation made within a Local Authority. Table 3.5 shows the number of appeals alongside the number of LEAs that operate the two different types of admission procedures. The ratio of appeals to LEAs is higher for LEAs operating open preference systems than for those operating the traditional catchment system. Forrest (1996) also shows that this is not simply an urban-rural dichotomy, by dividing the data into County LEAs and Metropolitan LEAs. The same pattern of high Appeals in LEAs with preference systems occurs irrespective of whether the LEA is relatively rural or urban.

Table 3.5   Number of appeals (1990-94) by admission procedures

	Catchment systems (*Preference systems*)		
	No. of Appeals	*No. of LEAs*	*Average per LEA*
County LEAs	15,385 (*9,918*)	31 (*9*)	496 (*1,102*)
Metropolitan LEAs	4,715 (*8,447*)	51 (*62*)	92 (*136*)

Source: Based on Forrest (1996)

Forrest (1996) proposes that there is greater parental dissatisfaction in Local Authorities that use the open preference system of admissions. However, it could simply be that those LEAs that operate these procedures are generating or encouraging greater market activity and therefore are inevitably going to have a greater absolute number of appeals. Also, the catchment system usually requires the greater amount of cultural capital to make a request for an alternative school since these LEAs require parents to write a letter to state their claim as to why the local catchment school is not desired. Paradoxically the LEAs that make parents apply directly to each school could be encouraging the most market activity and providing the most transparent system for the parents since they do not have to worry about making a strategic set of preferences. This is because only the parents know which schools they have placed a request for and so are more likely to be allocated to a school irrespective of which other schools they considered. Mayet (1997) identified that this kind of system constructs what was referred to as a 'power market' (p.170).

   It is difficult to identify if the choice of admissions system generates greater market activity since the decision to change the procedures for allocating places could be in response to greater market activity. Whichever way round this has come about the different admissions systems are inevitably going to produce different levels of market choice and competition. This might also be related to the length of time that elements of the market process have been working within a LEA. For example, just within the LEA-maintained school sector it was identified that even in the 1970s 27% of LEAs operated a system of parental choice (Dore and Flowerdew, 1978). Consequently it could be argued that schools in these authorities were exposed to competition much earlier than in other LEAs and that parents were also introduced to consumer empowerment at an early stage. This might help to explain why there is greater market activity in some Local Authorities than others.

More importantly these varying patterns of admission procedures are likely to be highly dependent upon the institution in charge and the approach the LEA takes to the education market as identified earlier by Ball *et al.* (1994).

Another key factor to the operation of the market place that is a product of institutional decisions by the LEA is the introduction of varying types of school provision over time. The next section deals solely with the levels of school diversity that have been generated over many years of policy-making by each LEA.

Diversity of Schooling

One of the underlying elements in the development of market processes in education has been the need for diversity within the choice of schools. Legislation throughout the 1980s and early 1990s has frequently been introduced to provide and encourage different types of schooling available to parents. Interestingly there has been little research that focuses on the diversity within education, particularly within state sector schooling. Glatter *et al.* (1997) proposed seven forms of school diversity: structural; curricular; style; religious/philosophical; gender; market specialisation; and by age range. Many of these forms can overlap to produce very unique sets of diversity in schools.

However, most policies throughout the history of state education have often promoted different **types** of schools rather than diversity within schools. And in order to understand the current landscape of school types it is necessary to take an historical perspective on education developments.

School development over time has been spatially uneven, both at a local scale and at a regional level. The first indication of this spatial imbalance could be seen in 1818 using the Brougham-inspired surveys (Marsden, 1986). Some of the northern counties were said to benefit from the very pro-schooling system in Scotland while urban areas tended to be the least developed. By 1851 there were considerable improvements across England, particularly in the Anglican heartlands, as Marsden (1986) called them, of the South East. But urban areas were still lagging behind and so too were the more remote counties of the South West and the Welsh borders.

Developments in the education system since the mid-19th Century have established and defined the forms of diversity currently in the market place. Since the first basis of mass education started with an uneven playing field it is not surprising that many of the developments have consequently also been spatially uneven, ironically to try and even out provision across the country. With different efforts to provide education has come different types of schooling and consequently varying spatial patterns of school diversity.

The Education Act of 1870 introduced School Boards, the first real state intervention, to try and fill the gaps in provision left by voluntary, often religious, organisations. As a result the first significant difference in schools opened up between schools sponsored and ran by the religious community and schools that belonged solely to the state. However, these School Board schools were not entirely free to the population. Fees were introduced at varying rates and as Marsden (1986) suggests these were fixed in order to keep schools socially segregated. Across the whole education system a social hierarchy of schooling was beginning to emerge with the upper classes sending their children to public schools, middle-class boys attending endowed grammar schools and working-class children having little option other than to attend the more affordable School Board elementary school (Ball 1986).

School Boards were replaced with LEAs in 1902 and were given powers to set up their own secondary (grammar) schools. At this time the state started paying for the education of those attending elementary schools and a system of scholarships were offered for grammar schools. However, it was not until the 1944 Education Act that secondary education was provided free for all people.

The 1944 Education Act continued the mass schooling based on a tripartite system, the first State attempt to purposively provide school diversity. The policy was to develop three types of schools that would meet different needs of children:

i.   Grammar Schools
ii.   Technical Schools
iii. Modern Schools

These different types of schools were meant to cross social-class boundaries but according to Ball (1986) middle-class children were consistently over-represented in grammar schools and working-class children were over-represented in modern schools.

After the Second World War the comprehensive movement began to gain pace by attempting to remove the diversity of school types. This approach to the education system largely came from the LEAs themselves but found opposition in Central Government, in particular the Labour Government of 1945-51. In 1951 there were just 13 schools that were deemed comprehensive but by 1961 this had increased to 138. However this was still short of the total 5,847 schools that already existed and in some form or other were of many different types of school. Most of these earlier comprehensive schools were in areas in which grammar schools would be unaffected (Ball, 1986). Therefore they were mainly built in new residential areas, such as New Towns and the large post-war Council Estates. As a result, if it can be said that grammar schools were located unevenly then so too were these early comprehensives.

In 1965 the Labour Government started to request that LEAs submitted plans for comprehensive reorganisation, but it was not until the 1976 Education Act that comprehensive schooling was formalised. Even with some obstacles to this change there was a dramatic shift from the tripartite system to a comprehensive system (Figure 3.8). It is between 1976 and 1979, towards the latter end of the growth in comprehensive schooling, that there was the clearest distinction between private and state education, particularly since Direct Grant Schools[2] were abolished and two-thirds became totally independent from the state (Bradford, 1993).

Clearly until the end of the 1970s there had been a fairly eclectic mix of schooling, thus reflecting different policy eras. For example, in the 19th Century most schools built were for the middle and upper classes and were consequently spatially located in the wealthy suburbs, as they were then, of towns and cities. Over time some new religious and endowed schools tried to provide for the less well off but they were still never usually in the poorest areas. School Boards were then given the responsibility to educate the working classes but as it has been seen these simply acted to perpetuate the social segregation of the cities. The next phase of large-scale school development was during the comprehensive era in education and many of the new schools were located in areas of new residential development, around the existing grammar schools. The last major development of this period then saw the reorganisation and conversion of many schools to comprehensive status. Some of these converted grammar schools will have retained their grammar ethos and reputation and so are still distinct from other schools. This was seen in the greater likelihood that ex-grammar schools would have sixth forms (Kerkchoff *et al.*, 1997). Even if they have structurally changed the perceived image of the schools as they were might still be evident. This final change is relatively recent hence many parents actually attended these older, usually grammar schools, themselves and will remember them as they were rather than as they are now.[3]

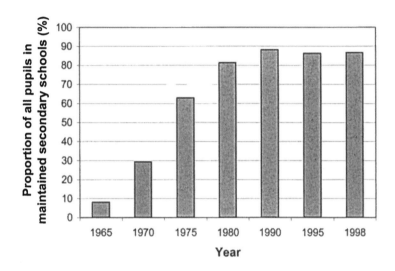

Figure 3.8 The growth of comprehensive education in England

This private-state distinction soon became blurred once again with the introduction of the Assisted Places Scheme in 1980. This was the first piece of legislation that turned away from the comprehensive approach and made grants available for bright pupils who could not afford otherwise to attend an independent school. In effect, just like the Direct Grant Schools, the state has once again begun to subsidise the private education sector. Further attempts by the 1979-97 Conservative Government to introduce diversity back into the education system have included the introduction of City Technology Schools, which are jointly financed with the private and the state sector, and Grant Maintained Schools, financed directly from Central Government, via the 1988 Education Reform Act. The 1993 Education Act provided a second push for Grant Maintained Schools along with the Specialist School Programme in which schools can specialise in a particular part of the curriculum and have a proportion of their intake based on ability in that specialism.

City Technology Colleges were still few in number by 1995 and had relatively little impact upon the market place. The first Grant Maintained (GM) school was established in September 1989 and by 1994 there were 814 secondary GM schools and 260 primary GM schools. It should be noted that since this research the status of Grant Maintained schools has changed; initially in the way they were funded and, more recently, their shift in status to Foundation schools, with greater LEA control.

It has been suggested that the introduction of these new types of schools have created a four-tier hierarchy within the English education system Burdett (1988, p.214) suggests that each tier has its own distinct geography and that the four-tier system 'is likely to be socially and spatially divisive.' However, even though this hierarchy exists based on levels of funding it is still unclear whether the hierarchy exists in the patterns of choice in the market place. For example, Bradford (1995) suggests that these tiers of provision do not necessarily produce hierarchies of outcomes once social class of the intakes has been taken into account. Therefore it would be premature to suggest that parents will choose one tier of education over another based on some hierarchy.

A more meaningful conceptual framework for considering diversity within the education market place is to use the **private-state continuum** developed by Bradford (1993). 'The changed state of the education system in England can be summed up at the national level as a blurring of the private/state divide' (Bradford, 1993, p.79). The **private-state continuum** attempts to distinguish between all schools in their level of 'privateness' (Figure 3.9).

This continuum does not necessarily assume there is a hierarchy between the different types of schools and also allows for there to be differences between schools of similar types. So, for example, Figure 3.9 shows there is a difference between LEA-maintained schools that are popular and those that are unpopular under market conditions. This would follow since popular schools are more likely to be oversubscribed and therefore are going to have to include some form of rules of selection in their admissions policy. The concept of a private-state continuum also allows consideration of other school features, irrespective of school type, that can affect the level of 'privateness', 'exclusivity' or 'control'. Therefore properties of schools that will impose upon the market place such as the gender policy of a school or the religious requirements of admissions can be included. This perspective on diversity within the education market is much closer to the theoretical approach to diversity that Glatter *et al.* (1997) proposed.

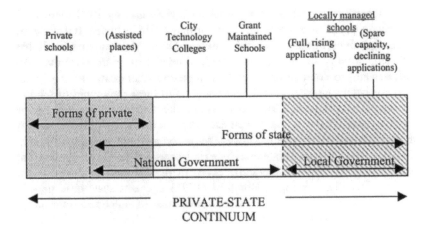

Figure 3.9   The 'private-state continuum' in educational provision

Source: From Bradford (1993, p.80)

Having now established a conceptual framework to consider diversity in the education market place the next section will examine the spatial nature of this diversity.

The Geography of School Diversity

It has already been seen that there are regional patterns in general education provision and outcomes but this Section will focus on the patterns of school diversity across the English education landscape.

Beginning with fee-paying or independent education a North-South divide was first recognised by Coates and Rawstron in the 1960s (1971). The spatial imbalance in fee-paying education generally follows the distribution of wealth across England, but Bradford and Burdett (1989) observed that even allowing for social class differences there was still a greater propensity for private education to be in particular areas, such as the south-east. These same authors have also shown that this North-South divide in fee-paying education still exists almost two decades later and that it had grown. The situation in 1996 is very similar, but of particular significance is the presence of fee-paying education in the South-West (Figure 3.10).

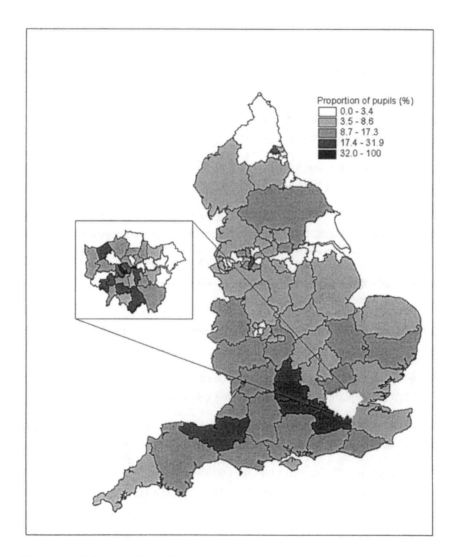

Figure 3.10 Proportion of secondary school age pupils in fee-paying schools, England January 1996

As mentioned in the previous section the distinction between fee-paying and state education has become less obvious over the last two decades. However, the emerging private-state continuum has not developed equally across the country.

'While this continuum is characteristic of the national situation, however, there is much variation at the local level' (Bradford, 1993, p.80). Bradford (1995) has also illustrated the introduction of GM schools within the context of existing fee-paying education. First of all, GM uptake has been spatially marked (see Figure 3.12) because most of the early conversions to GM status were in Conservative controlled LEAs, and were schools that were already selective or single-sex or those that were threatened with closure. Therefore, Bradford (1995) distinguished between Local Authorities that had high levels of fee-paying education but with little development of GM schools and Local Authorities that had low levels of fee-paying education but have had a high conversion of schools to GM status. The resulting four sets of LEAs showed that the continuum had not developed evenly across England.

The development of a private-state continuum in each LEA also needs to address other forms of private-state schooling such as religious schools. Figures 3.11 to 3.14 show, for these LEAs, the proportion of schools that were comprehensive, GM, fee-paying and 'other' (i.e. voluntary controlled or voluntary assisted) schools in 1998. Unfortunately comparison is limited because a well-developed and evenly developed private-state continuum would have a low proportion of all four types of schooling. Therefore, to take into account the variations in the private-state continuums for each LEA an analysis of the distribution of these different schools must be made. Taking a cluster analysis using the proportion of schools as either comprehensive, GM, religious, or fee-paying from each LEA, six different forms of private-state continuums across LEAs can be seen. The proportion of schools under different levels of private control and the resulting levels of school diversity are summarised in Table 3.6. Each set of LEAs from this cluster analysis represents a different form of the private-state continuum between a 'private and well-developed' continuum to a 'state' continuum.

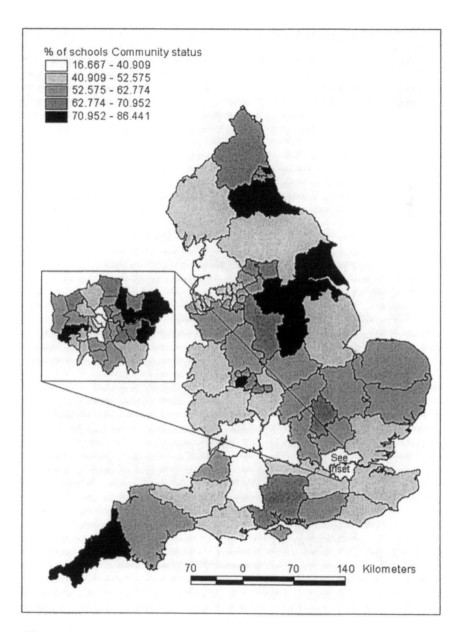

Figure 3.11 Proportion of schools by LEA that were comprehensive schools (community status), England 1998

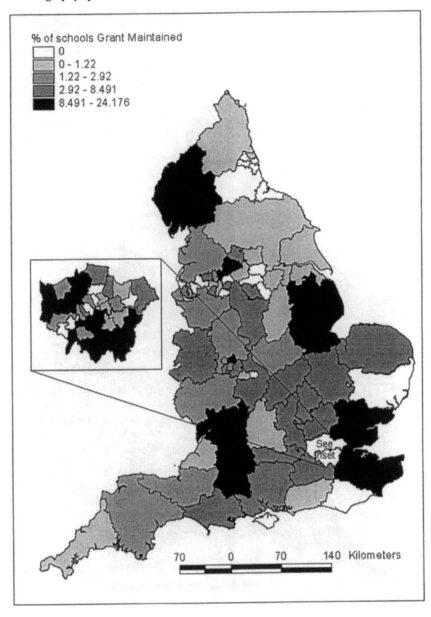

% of schools Grant Maintained
- 0
- 0 - 1.22
- 1.22 - 2.92
- 2.92 - 8.491
- 8.491 - 24.176

See inset

70   0   70   140 Kilometers

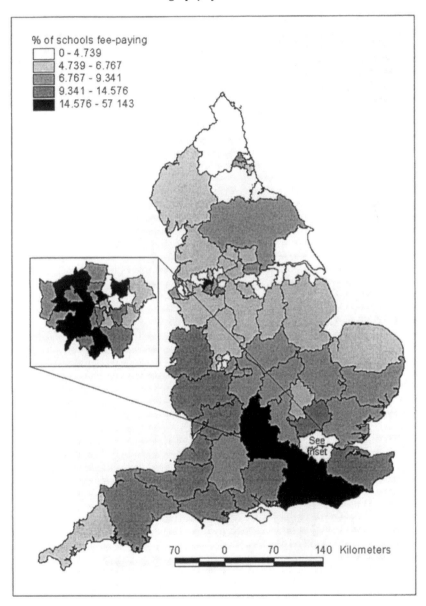

Figure 3.13 Proportion of schools by LEA that were fee-paying, England 1998

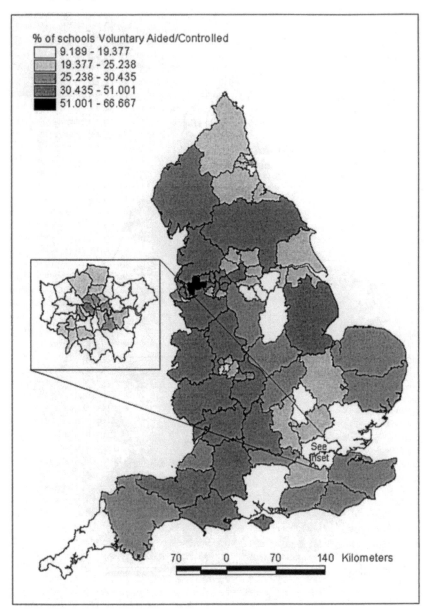

% of schools Voluntary Aided/Controlled
- 9.189 - 19.377
- 19.377 - 25.238
- 25.238 - 30.435
- 30.435 - 51.001
- 51.001 - 66.667

See inset

70   0   70   140 Kilometers

**Figure 3.14 Proportion of schools by LEA that were voluntary aided or
voluntary controlled, England 1998**

Table 3.6 Six local private-state continua, based on school composition of LEAs

Local Private-State Continuums	Description	Approximate proportion of schools by LEA, 1998			
		Comprehensive (%)	Grant Maintained (%)	Fee-paying (%)	Voluntary Aided or Controlled (%)
P-S 1 – Private & well-developed continuum	The most complete private-state continuum, with all types of schooling represented fairly equally.	22-32	0-3	33-37	30-44
P-S 2 – Semi-private & well-developed continuum	Fairly well-developed private-state continuum but lacking a significant presence of religious schools.	38-55	0-12	14-33	19-30
P-S 3 – Religious & well-developed continuum	Fairly well-developed private-state continuum, but, unlike Cluster 2, has a greater presence of religious schools at expense of fee-paying sector.	38-60	0-16	0-16	33-52
P-S 4 – Religious & un-developed continuum	No real private development of schooling but a very significant presence of religious schools.	16-37	0	0-17	63-67

Table 3.6   (Continued)

P-S 5 – State-private continuum	Even development of fee-paying and religious schools, but with relatively large Community school presence.	47-63	0-24	5-20	12-31
P-S 6 – State continuum	Mainly state schools but there is some private presence, usually *either* religious *or* fee-paying schools.	62-87	0-10	0-19	9-32

Figure 3.15 shows the spatial distribution of these clusters of private-state continua. The first observation of this analysis of LEA school composition is the relatively insignificant presence of GM schools on the development of the local private-state continua. For example, in the LEAs that could be classed as exhibiting a well-developed private-state continuum there was almost no incidence of Grant Maintained schools. Indeed, there was no particular local private-state continuum identified here that was more likely to have GM schools present. This would suggest that the propensity of schools to become GM in England had little or no relationship with the degree of privatisation, within the respective LEAs, already in place.

The most 'private' and well-developed private-state continua occur in Inner London extending to the West of Outer London and into parts of the adjoining Counties of the South. The significance of more religious but still relatively well-developed private-state continua can also be seen, typically in county LEAs of the West and metropolitan boroughs of the North West. If the presence of religious schools is ignored for the moment then there is a distinct north-south divide emerging, with the South East generally having the greater levels of 'private' development in their education systems. The most 'state' and un-developed 'private' systems are found in the West Midlands and the North East of England.

This classification is only based on school types and does not include other features, as discussed earlier, that can make a school private. There are three different forms of exclusivity or privateness of a school and these are summarised in the following table (Table 3.7).

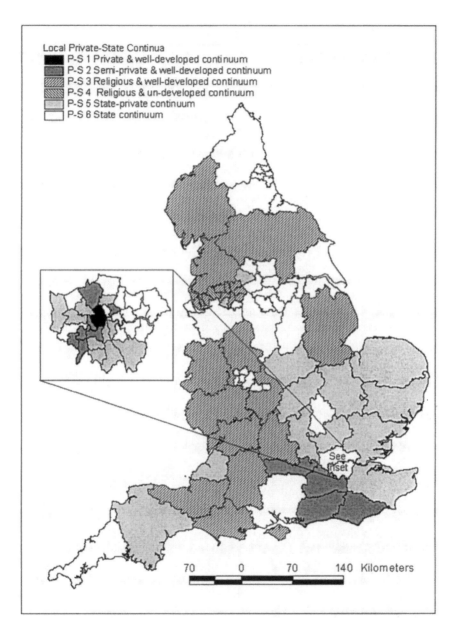

Figure 3.15 Distribution of local private-state continua in England, as at 1998

Table 3.7   Forms of 'privateness' in schools

Elements of private-state continuum	School characteristics
Sex policy	Mixed, Girls or Boys
General admission policy	Comprehensive, Modern, Fee-paying or Selective
Control of school	LEA-maintained school Voluntary Controlled school Voluntary Aided or Special Agreement Grant Maintained Independent

Each of the elements to a private-state continuum have their own degree of privateness and hence a school can have any combination of these characteristics that makes them unique along the private-state continuum. Using a numeric score for these school characteristics allows all the schools to be aggregated in a Local Authority to produce a private-state index (Figure 3.16). The different scores for each school characteristic ensure they are weighted according to their degree of 'privateness' or exclusion.

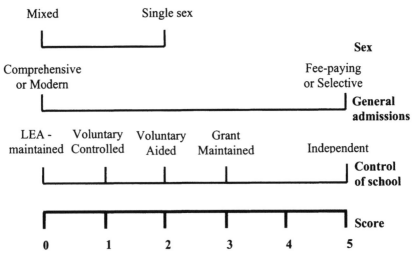

Figure 3.16   Scores for private-state characteristics

It should be noted that these scores only comprise individual elements of a school. By summing the scores for each school characteristic a total figure is produced, the lower the figure the closer the school is towards the state end of the private-state continuum and the larger the figure the more private the school. So, for example, a fee-paying independent school will have a private-state score of 10 (fee-paying (admissions) = 5; and independent (control) = 5). If this private school were also a single sex school then the private-state score would be 12, which incidentally is the highest private-state score achievable.

The school private-state scores can then be aggregated to produce an index score for the whole Local Authority. Each school's private-state score needs to be weighted by the number of students attending that particular school in order to get an accurate indication of private-state education **consumption**.

Figure 3.17 illustrates the spatial distribution for selected LEAs of the level of private and state education across England using the private-state index. Again the South and the South East tend to have the greatest degree of private education, the exception being inner parts of London. Other metropolitan areas outside London, such as Birmingham and Walsall, also feature a good degree of education privatisation. Three remote county LEAs also appear to have a high level of private education, Cumbria, Lincolnshire and Shropshire. Local Authorities with a relatively low level of private education are Suffolk, Somerset and Cornwall. Interestingly the latter two counties have a high proportion of pupils attending fee-paying schools but clearly there has been little privatisation of the state education sector.

In order to see if the diversity of schools impacts upon the market place it is necessary to examine the level of parental willingness and empowerment to engage with the market place at this scale.

The Geography of 'Appeals'

The analysis has now considered the policy background to changes in the landscape of education and the resulting diversity of choice in each market place. It is still unclear as to how the level of diversity will impact upon the market place and, in particular, the choice that parents will make as to which school to send their child to. For example, is it possible to suggest that the greatest amount of market activity will take place in areas with a high level of private-state education? As Bradford (1990) proposed, this could work in two opposite ways. Greater market activity could occur

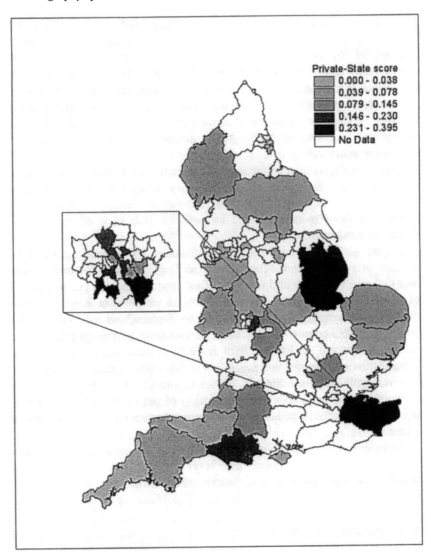

Figure 3.17 Aggregate private-state scores for a sample of LEAs (LEA-maintained schools only, weighted by number of pupils), England 1994-95

where there is greater diversity and privatisation of schooling because an element of private consumption and competition between schools already exists. On the other hand there may be more parents willing to engage in the market in areas of low fee-paying schooling because in these areas the more able, both culturally and materially, are still in the state sector.

Since it is nearly impossible to generate levels of market activity for all the English Local Authorities a useful way of examining the nature of parental empowerment in the market place is the number of appeals lodged by parents in each LEA. If a parent is dissatisfied with the school place allocated to their child even after their preference has been considered the LEA has to give them the right to appeal against the decision. First of all the appeal is lodged, then if the parent still wishes to go through with the procedure the appeal is heard by an appeals committee. This committee then decides if the appeal should be rejected or if it is successful then the initial allocation of a school place will be changed.

The total number of appeals has increased year on year. On the basis of the Statistical First Release SFR 19/99 the admission appeals lodged for maintained primary and secondary schools are summarised in Table 3.8.

Table 3.8   Total number of appeals lodged for maintained primary and secondary schools, 1995-96 to 1997-98

Year	Number of Appeals
1997-98	77,000
1996-97	72,700
1995-96	62,900
1993-94	34,740
1988-89	18,040

The total number of appeals lodged for 1997-98 represents a 22% increase over the previous two years but is also four times more than it was in 1988-89. However, these appeals are not distributed equally across the LEAs. Figure 3.18 shows the spatial distribution of secondary school Appeals lodged as a proportion of secondary school age pupils for 1996-97.

The LEAs with a high occurrence of appeals are typically urban/metropolitan authorities. However, it should be noted that this is not the case for all metropolitan Authorities. Other county LEAs with a high level of appeals lodged are Lancashire, Avon, Shropshire, Warwickshire, North Yorkshire, Humberside and nearly all of the home counties.

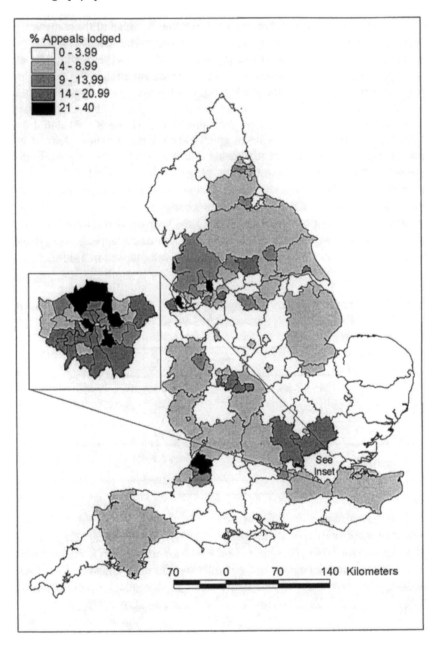

Figure 3.18 Appeals lodged as a percentage of LEA admissions, England 1997-98

The greater propensity for appeals to be lodged in urban areas is probably because these authorities are the more active market places simply because they offer the greatest amount of school choice.

Overriding this is the variation in admissions procedures that, as Forrest (1996) identified, may generate differences in parental satisfaction. Forrest (1996) suggested that LEAs who operate open preference systems of admissions were more likely to have appeals lodged against the final allocation of places. But as highlighted earlier these kinds of admission procedures could be generating greater market activity that is inevitably going to produce a greater number of parents wishing to appeal over their allocated school.

The following Table 3.9 shows the significant relationships with the number of Appeals lodged for 39 LEAs. The urban-rural contrast of a Local Authority is still as significant as it was for **all** the English LEAs. But what also appears to be important is the education expenditure per pupil (SSA) and the private-state index. However, the expenditure variable was not as significant for **all** the Local Authorities and probably only features because there is a strong relationship between expenditure and the urban-rural nature of a LEA. But it is clear that the extent of privatisation in the market place, even just of the state sector, might have some bearing on the level of appeals lodged in the market place.

Table 3.9 **Significant correlations with appeals lodged for a sample of LEAs (n=39)**

Variable	Non-parametric correlation (R)*
Density of schools	0.641
Density of students	0.639
Standard Spending Assessment (SSA)	0.621
Private-state index	0.594
Private-state index (state schools only)	0.571

* Correlations are significant at the 0.01 level (2-tailed)

The number of appeals lodged, whether representing consumer empowerment or dissatisfaction, is likely to be related to the density of schools, urban or rural, and the status of the private-state continuum in a Local Authority. Using linear regression it is possible to see how all the characteristics of a LEA might influence the number of appeals lodged.

Regression Analysis of the Number of Lodged Appeals

In order to identify the key factors that may have determined the number of appeals lodged a regression model was developed. (Appendix A lists all the LEA variables available for the regression analysis to predict the number of appeals in each Local Authority.)

The regression model developed to predict the number of appeals lodged in a LEA used backward linear regression for the sample of LEAs (n = 39) but with a pre-selected set of variables as predictors of the dependent 'Appeals lodged'. The purpose of this model was to develop a model that was relatively significant but that used as few variables as possible. The variables used in this model were as follows.

Variables entered:
Standard deviation of GCSE results within LEA
Density of schools
Proportion of failing schools
Proportion of surplus places
GCSE results (% obtaining 5+ GCSEs grades A-C)
Standard deviation of private-state index within LEA
Standard spending assessment (SSA)

This produced a model using the 7 pre-defined variables (Table 3.11) and an Adjusted $R^2 = 0.690$ (Table 3.10).

Table 3.10 Model summary of backward regression with selected variables

R	R^2	Adjusted R^2	Standard Error of the Estimate
0.864	0.747	0.690	0.2522

Table 3.11 Coefficients of backward regression with selected variables

Variables	Unstandardised Coefficients		T	Significance
	B	Std. Error		
Constant	-1.887	0.536	-3.518	0.001
Density of schools	5.184E-09	0.000	5.082	0.000
GCSE results	3.135E-02	0.006	4.869	0.000
Stdev. GCSE	3.322	1.107	3.000	0.005
Failing schools	9.639E-03	0.004	2.480	0.019
Stdev. private-state	-1.950E-04	0.000	-2.137	0.040
SSA	2.978E-04	0.000	1.843	0.075
Surplus places	-1.543E-02	0.015	-1.045	0.304

The formula for calculating this model was as follows.

Regression equation:

Appeals = -1.887 + {density of schools}*5.184E-09 + {GCSE}*3.135E-02 + {stdev. GCSE}*3.322 + {failing}*9.639E-03 - {stdev. private-state} *1.950E-04 + {SSA}*2.978E-04 - {surplus}*1.543E-02

This model was fairly significant considering that it did not include the different admission procedures that applied to each of the Local Authorities in this analysis. Obviously using more variables would have produced more significant results. What is clear from these calculations is that the variables acted in the way one would expect them to influence the number of appeals lodged. The density of schools and the private-state index have already been discussed, but it is not surprising that with a greater number of failing schools there are a greater number of appeals with parents trying to avoid those failing schools. The greater the variation in GCSE results the greater the perceived differences between schools and therefore the greater need to avoid particular schools. The fewer the number of surplus places within the market place the harder it would be to get a school of a parent's choice and again therefore the greater number of appeals. Finally the higher the level of deprivation, based on the Additional Education Needs (AEN) calculated by the DfEE, within a LEA the greater the need to avoid the 'sink' schools that are likely to be associated with poor areas.

Of all these variables and the possible conclusions one can draw from them it still remains that the private-state continuum, either through the index score or the degree of variation within the Local Authority, continues to be of importance.

Market Activity and Appeals

In order to examine the relationship between the levels of market activity and the number of appeals this section will focus on the LEA variables for eight Local Authorities with admissions data.[4]

Figure 3.19 shows that there was a relationship between the number of appeals and the level of market activity as defined by the number of students attending an alternative school to their nearest school (non-parametric correlation of R = 0.857).

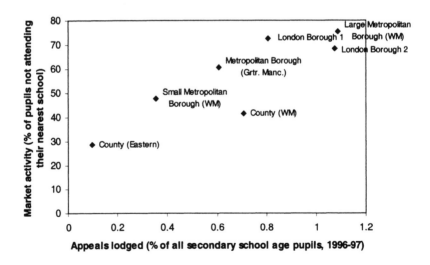

Figure 3.19 Relationship between market activity and appeals lodged

However, appeals do not necessarily mean that there is greater market activity across the Local Authority. Appeals are also a sign of dissatisfaction with the process, largely because there are very popular but over-subscribed schools within the LEA. This situation can create 'hot-spots' within a locale (Audit Commission, 1996).

It would appear that the number of appeals is dependent upon a combination of both market activity and factors inhibiting parental preference. Ultimately it is also an indication of parental empowerment within the market place.

Studying the relationship between the LEA variables used above and the level of market activity and number appeals shows some interesting outcomes (Table 3.12).

Table 3.12 Relationships with appeals lodged and market activity

LEA Variables	Non-parametric correlations	
	Appeals	*Market activity*
Density of schools	0.667	0.881[b]
Standard spending assessment (SSA)	0.762[a]	0.810[a]
Pupil-teacher ratio	-0.719[a]	-0.695
AEN – Index of deprivation	0.619	0.833[a]
Private-state index	0.714[a]	0.238
Private-state index (state)	0.762[a]	0.357
% Comprehensive schools	-0.690	-0.238

a   Correlation is significant at the 0.05 level (2-tailed)
b   Correlation is significant at the 0.01 level (2-tailed)

There were three variables that have a significant relationship with both the number of appeals lodged and the level of activity in the market place, density of schools, education expenditure per pupil and the pupil-teacher ratios. It is likely that the expenditure and pupil-teacher ratios are, themselves, related (R = -0.683), and that the expenditure in LEAs reflects the urban-rural contrast (R = 0.881).

The DfEE's own index of deprivation also shows a positive correlation with appeals lodged and market activity. However, there is a much stronger and more significant relationship between the AEN and market activity. This is the first indication that there are differences in the characteristics between LEAs with similar number of lodged appeals. It is possible to suggest, therefore, that the urban deprivation of these eight Local Authorities is linked to higher market activity but this does not necessarily mean that these same LEAs are going to have a relatively high number of appeals. Instead it is the Local Authorities that have a high private-state index that are more likely to have a high number of appeals lodged against the allocated places.

This would indicate that parents living in urban Local Authorities, and in particular, the more disadvantaged urban environments, are more likely to be engaging with the market place than those parents living in rural areas. However, of these active market places there are higher levels of dissatisfaction in areas where there has been greater education privatisation. This could be because with greater private education there is greater exclusion and selectivity that could lead to greater frustration among the parents. An alternative explanation might suggest that the private-state continuum is perceived to be hierarchical by parents and therefore generates greater keenness, and consequently greater market ferocity, to get a place in particular schools.

Conclusions

It has been seen that the number of appeals lodged has some relationship with the levels of activity in the market place for the eight LEAs with admissions data. There is also a significant relationship between appeals and the private-state continuum.

The private-state continuum could influence the level of choice in a number of ways. The first is that greater choice based on the diversity of schools, and the resulting competition that is generated, has historically embedded the choice process of comparing and evaluating schools, even before many of the present reforms took place. The second way that the private-state continuum could influence choice in the market place is that with greater private education then the levels of private consumption could also be higher. For instance, this might have generated a consumer ethos that makes it acceptable to reject the local school even if it is in favour of another state school.

There has been very little written about private consumption of traditional public services (see Clarke and Bradford, 1998) and it has been said that it is becoming harder to distinguish between public and private consumption (Dowding and Dunleavy, 1996). But whatever defines this 'private' consumption 'the educational consumption cleavage is becoming more spatially marked' (Bradford, 1989, p.153). Such cleavages across England in private consumption cannot be simply explained by social class. Obviously levels of income do affect the amount of private consumption occurring but there are still disproportionate amounts of this private consumption in some areas, even when accounting for social class (Bradford and Burdett, 1989).

There are two sides to explaining different levels of private consumption (Bradford and Burdett, 1989). The first are supply-led factors, i.e. since there is greater private provision of a particular service then more people will consume these services. The examples used in this study do not necessarily show that there has to be a considerable level of private supply for parents to engage more with the market. Hence the other set of explanations might be of some significance. These are demand-led factors in which there is a stronger culture of private consumption brought about by, for example, a desire for people to increase the control they have over their lives (Saunders, 1990). These strong individualistic needs are then socialised in space via the neighbourhood effect (Bradford and Burdett, 1989).

This private consumption of state services does not only affect education. Pinch (1997) has agreed that a lot of the variation in private ownership of council houses under the right-to-buy legislation of the 1980s is not entirely explained by social class. Similarly Mohan (1995) acknowledged the very high provision of private health care in the South and South East.

This would suggest that another element to the education landscape that will affect the consumption in the education market is the varying levels of private consumption, the culture of individualism and non-reliance on the state as the only provider of traditional welfare services. It has been shown that the spatial variations in the private consumption of a number public goods is similar to the propensity for parents to lodge an appeal over the admissions process (Taylor, 2000). These included the private consumption of council houses, medical insurance and private pensions.

These three examples of private consumption can be related to the levels of appeals lodged in education (Figure 3.18). In all the cases the South and South East feature quite strongly as areas of high private consumption. Conversely the levels of private consumption in the East are much lower in each of the examples than perhaps expected. Also the number of appeals lodged tends to be higher for LEAs in the West Midlands than those in the East Midlands, which is similar to the patterns of private consumption of pensions and council houses. The only area where the number of appeals is higher than one would expect having considered the other examples of private consumption is in the North West. However, the North West contains a large number of public schools, such as the old Direct Grant Schools, therefore having a history of private consumption in education. The North West also has a higher private consumption than the North East region.

This does not suggest that it is the same people who consume these 'private' services, but that these different areas might represent varying cultural climates in the propensity and empowerment to consume private services. Consequently this shows that the propensity to consume private services generally could have a bearing upon the levels of activity and parental empowerment in the education market place. However, it is impossible to ignore the level of private and diverse provision in the market place, the private-state continuum.

Overall this discussion has shown that a number of factors will have an effect upon the processes at work in the education market:

- LEA policy – historically has determined diversity and competition.
- LEA policy – admissions procedures affect levels of engagement with the market.
- Spatial distribution of schools and consumers – access to choice.
- Diversity of provision, in particular the private-state continuum.
- Socio-economic deprivation and the need to be active in the market.
- Levels of private consumption.

Notes

1   Points generated from school postcodes were used to create thiessen polygons in ArcInfo (GIS) as surrogate catchment areas. Apportioning points into regions creates thiessen polygons, otherwise known as Voronoi or Proximal polygons, so that any location with these regions would always be closer to the region's point than to the point of any other region. Consequently, they can be used to delimit pupils to their nearest school in each LEA. The resulting areas of these 'thiessen' catchment helps indicate the urban-rural nature of schools within an LEA.

2   Direct Grant Schools were financed by a mixture of endowment, fees and state grants conditional upon admittance of a percentage of non-paying pupils.

3   An example of the significance of this is in Leicestershire, the first LEA to remove all selection from schools, i.e. the comprehensive ideal. But of the fifteen county upper

secondary schools twelve were formally grammar schools and some have retained their grammar title.
4   The LEAs with admissions data are: Inner London borough (Eastern), Inner London borough (Western), Outer London borough (Northern), large metropolitan borough (West Midlands), small metropolitan borough (West Midlands), metropolitan borough (Greater Manchester), county (Eastern), and county (West Midlands).

4 The 'Lived' Market Place

Introduction

In Chapter 1 the concept of the **market place** was defined as the physical locale in which most schools compete with one another and where the majority of children attend schools. It was consequently argued that the most obvious examples of market places were spatially defined at the LEA level, and included the schools from the state education sector and the parents and children living within these boundaries. However, the notion of the market place is only useful in providing a framework, both conceptually and geographically, to study, and understand what Waslander and Thrupp (1995) called "the 'lived' market". To understand how the market operates it is necessary to examine the patterns and processes of competition occurring within the market place.

This Chapter is the first of four that examine the processes and dynamics of the education market in reality. This discussion of the 'lived' market place fits within the context outlined in the previous Chapter by shifting down in its spatial level to the *school scale*. It must be remembered that the market operates at a variety of spatial scales. However, since we are taking a more geographical perspective of the market the school scale is probably the best level in which to consider real space and the locations of schools and children within a Local Authority.

By analysing admissions data of 34,178 pupils attending 198 LEA-maintained secondary schools in eight LEAs[1] this Chapter begins by discussing some of the real occurrences of the market place within different geographical market places. It attempts to understand the complex patterns of school choice under market conditions and begins to define the competition occurring within the market place. This is achieved by considering three elements of the 'lived' market place. The first is to define the actual areas of competition between schools, i.e. **competition spaces**, and identifies the different features that can constrain and control competition in these competition spaces. From examining admissions data for all state schools within the eight LEAs the second part of the discussion identifies and defines the different ways in which the education market operates. In other words, it highlights how schools compete against one another. The final element to the 'lived' market place discussed here argues that competition spaces vary within and across market places. However, it also suggests that they can be categorised into four different types of competition spaces and identifies their important features. This analysis and discussion of the 'lived' market place provides empirical evidence and, therefore, a greater understanding, of how the 'new' education market operates.

Competition Spaces

The market place can be a very complex arena of patterns, processes and decisions. The resulting outcome of the education market can be just as complex. For example, do all schools compete with one another? Do the schools simply aim to serve their immediate locale and community? Or do the schools compete across an area that could be defined as some sort of sub-region?

Because of this complexity the market place needs breaking down in to what can be termed **competition spaces**. This is best viewed from the school-scale perspective and thus considers the location of schools and pupils alongside the resulting market performance of the schools. These competition spaces could be described in terms of whether the competition within the market place is on a level or uneven playing field. This is largely determined by the spatial distribution of schools and pupils, and differs between and within LEAs. Consequently, the competition space can be defined as the area of interaction between schools in the market place. These areas can vary in size and character and can be measured in many different ways. It is possible to have several competition spaces within a LEA, or the LEA may be a single competition space in itself, if not a very complex one. It is also possible that the competition spaces overlap or form a hierarchical market place. Attempts at defining the real occurrences of such competition spaces will help to answer questions raised above about who competes with whom.

A useful position from which to start is by considering Glatter and Wood's (1996) 'local competitive arenas'. These form a basic spatial model of the market. It could be argued that the 'local' refers to a particular competition space and at a particular scale. Each 'local competitive arena' fits with other 'local competitive arenas' to form a mosaic of such competition spaces across the LEA. However, this basic model works on the basis of proximity, i.e. adjacent schools competing against one another for pupils living within just their immediate vicinity or catchment. It is likely that for the majority of schools this may be the case but there are a good number of examples of competition spaces that cannot be defined like this and are not based simply on schools being adjacent to each other. This could be the case, in particular, for the more 'private' schools, such as religious schools. These would not necessarily be competing against other neighbouring schools but, instead, with other similar religious schools, wherever they are located within the LEA.

Glatter and Wood's 'local competitive arena' however, does provide a good starting point from which to construct these competition spaces that comprise the market place. This Chapter will examine whether it is possible to identify such patterns from admissions data by looking at competition spaces in reality.

In order to examine the competition spaces of schools in the education market it is necessary to consider first the spatial landscape that provides the foundation to the competition space. Consequently three important geographical characteristics need to be identified:

i.   Proximity to other schools.
ii.   Urban-rural context.
iii.   Accessibility.

Proximity to Other Schools

Glatter and Wood's (1996) model used, to some extent, the proximity of schools to determine the 'local competitive arena'. It is not clear if the degree of interaction between neighbouring schools is a product of the distance between schools, or if it assumes that there is equal distance between schools. The 'local competitive arena' does suggest that there is a higher degree of interaction between adjacent schools as compared to non-adjacent schools. This supports the claim that proximity is being used to define the competition space.

To illustrate the variation in proximal distances between schools within each LEA, thiessen polygons, i.e. surrogate catchment areas, were generated in ArcInfo (GIS) around each school. The resulting polygons can be viewed as representing the 'local' catchment of a school. The zoning of these areas allocates pupils to their nearest school. Some schools have relatively smaller thiessen polygons, or local catchments, than others, even within an urban LEA. This reflects the uneven distribution of schools within a LEA. This unevenness of the size of local catchments can make some schools more susceptible than others to the market and its forces. But this does not mean that they will lose out in the market place. Schools close to each other can be either more susceptible to losing their local catchment pupils or more likely to attract pupils from the other schools' local catchments.

Urban-Rural Nature of School Catchments

The majority of research on UK education markets has focussed almost entirely upon urban areas. However, many LEAs and schools are not completely urban or do not have an evenly distributed population or set of consumers. Hence competition spaces may exist in different forms because they are urban, rural, or even suburban. The competition spaces may even overlap these geographical characteristics and produce uneven 'playing fields' as a result.

An attempt to incorporate the varying degrees of proximity between schools has been considered before by the Audit Commission (1996) in its report on the allocation of school places. Five different types of LEAs were identified:

i.	London Boroughs	33	(31%)
ii.	Metropolitan Districts	36	(33%)
iii.	Urban Counties	7	(6%)
iv.	Semi-rural Counties	18	(17%)
v.	Rural Counties	14	(13%)

Using this classification it is clear that there is a significant proportion of LEAs across England that are rural or semi-rural in nature (30%). This makes it even more important that schools in a rural context are included in such analysis of the quasi-market in education. However, a simple classification of the Local Authorities may hide key differences within the LEA. For example, it is quite likely that within the Rural Counties there are some urban schools that have urban catchments (see previous Chapter for further discussion and illustration of this).

A basic way of identifying geographic differences is to use the area size of the thiessen catchment areas generated from the school locations. Cluster analysis of the resulting school catchment sizes suggests that there were eight significant catchment types ranging from very urban to very rural (Table 4.1).

Classifying the size of catchment areas to a general urban-rural continuum is very important since there are likely to be many differences in the way the market operates in these different geographical areas. The issue of proximity between schools identified earlier will have varying impacts on urban or rural schools. The proximity of schools in rural areas can be beyond a particular threshold in which the market processes and dynamics are significantly different to what would normally be expected. It is often assumed that rural areas have no 'parental choice'. However, the amount of parental choice might be less in absolute terms but the processes operating within the market place could be as significant in the way they work as in the urban market place. This urban-rural variation should also not exclude the more suburban schools. Again, the way the market functions in these areas could be very different simply because the socio-economic characteristics of the intakes are likely to be significantly different. Also, there are probably more constraints on choice than in the urban market place but there is likely to be more available choice than in rural areas.

Table 4.1   Catchment area classification

Catchment type	Area (metres)*	Proportion of schools in study (%)
Urban1	4,500 – 16,999	71.2
Urban2	17,000 – 49,999	7.6
Suburban1	50,000 – 79,999	7.6
Suburban2	80,000 – 109,999	3.3
Rural1	110,000 – 144,999	2.2
Rural2	145,000 – 174,999	3.3
Rural3	175,000 – 234,999	3.3
Rural4	235,000 – 324,999	1.6

* Range as defined from cluster analysis of all thiessen polygon sizes.

Accessibility

The physical accessibility of a school to parents and their children is a key constraint on their choice of schools. This is not necessarily related to proximity but one can assume that schools within close proximity (urban classification) are more accessible to more people than those schools that are further apart (rural classification). As mentioned earlier accessibility can work in two ways for a school. It can allow a school to attract pupils from another catchment area but it can also make a school susceptible to losing pupils to a competitor.

Accessibility within education provision can come in many forms because children travel to school in a number of different ways. For example, some pupils travel to school under their own means of transport, such as by walking or cycling. For these pupils a 'safe', and relatively short, route is desired to school. Other pupils rely on their family's means of transport, such as by car. The constraints on this form of travel depend on the flexibility in the use of the family car. For example, does the car have other objectives such as getting a parent to work? If so, then the route to a school has to generally fit around these other objectives. The flexibility of car use is also determined by the availability of more than one car or flexible employment conditions, of which either could allow more time to travel to alternative schools. A third set of transport options use independent means of transport. These can come in the form of a school organised bus service, the use of public transport, or in some instances a community bus service organised by the parents in a particular locale. These options depend on the particular routes available to the pupils, the cost of using such transport (it should be noted that not all school organised bus services are free), and in the case of the last option, the necessary time and inclination of parents to organise their own bus service.

The issue of how to get to school is a very complex one because these decisions are unique to each household unit. Such decisions are examined in more detail in Chapter 6, but for the purposes of the analysis on the market place in this Chapter proximity was used as the only function of accessibility. Consequently, the actual location of pupils becomes the most important factor in this context, and as discussed above these locations are unevenly distributed and change every year.

Patterns of Competition

When defining the competition space it was suggested that schools could compete with each other or they could compete in smaller sub-areas, such as the 'local competitive arena'. The question now arises whether such a pattern of competition can be identified.

One possible way of identifying which schools compete with each other is to ask the schools themselves. However, as discussed earlier, the notion of competition can be regarded as a fairly negative and somewhat powerful concept. Therefore, any kind of indirect competition will often be overlooked within the response. Since most schools are not necessarily operating under market principles themselves there is a good chance that very little acknowledgement of competition will be given. However, it is fairly clear that the outcomes of the new legislation are market-based. The consumption of education, and giving parents 'choice', drives the market – i.e. demand-led competition. Therefore, it is possible to identify the patterns of competition from a post-choice, or demand-led, perspective. It is the results of the market that can inform us about the competition spaces.

If it is assumed that without the quasi-market in education then nearly all pupils would have attended their **nearest** school then any deviation away from this can be regarded as the outcome of the market process. Therefore, it can be said that competition exists between two schools if one of those schools attracts pupils from the other schools' 'local' pool of pupils. Using this approach for all schools within a Local Authority it is possible to generate a matrix of 'gains' and 'losses' between the schools. Figure 4.1 illustrates how a matrix of gains and losses between schools can be generated, and Figure 4.2 provides an example of such a matrix from an LEA in London.

	School 1	School 2	School 3	.....	School n
School 1	L_1	G_{12}	G_{13}	.....	G_{1n}
School 2	G_{21}	L_2	G_{23}	.....	G_{2n}
School 3	G_{31}	G_{32}	L_3	.....	G_{3n}
⋮	⋮	⋮	⋮	⋮	⋮
School n	G_{n1}	G_{n2}	G_{n3}	.....	L_n

L_i = Number of 'local' pupils attending school i
G_{ij} = Number of gains to school i from school j
G_{ji} = Number of gains to school j from school i
n = The total number of schools in a LEA

This suggests that the total number of possible 'events' occurring (i.e. the total number of gains **and** losses) within a LEA is = $(n \times n) - n$.

Figure 4.1 Matrix of gains and losses in a hypothetical LEA

	Schools							
	chr	cop	eas	edg	fai	que	rav	whi
chr	69	17	1	0	21	0	1	24
cop	23	106	0	20	0	0	0	24
eas	2	0	101	0	19	7	41	1
edg	0	60	0	69	0	0	0	15
fai	21	11	8	3	82	0	7	3
que	21	2	20	0	52	29	45	0
rav	4	6	7	3	15	26	35	0
whi	2	8	0	1	0	0	0	77

Figure 4.2 Matrix of gains and losses between schools in an outer London borough (Northern) (1995)

From these matrices it is possible to identify which schools appear to compete with each other. The matrix of gains and losses is an 'abstract' space and does not really illustrate the **spatial patterns** of competition. However, the actual spatial flows of students as gains and losses are so complex that these matrices provide a useful tool in order to begin analysing competition (Taylor, 2000).

It is still the case that the patterns these matrices produce are very complex. Therefore, it is very difficult to make a single interpretation of such competition. In particular it is not totally apparent that the 'local competitive arena' exists in the form that Woods *et al.* (1996) suggest, whether in urban or rural areas. In order to extract a useful understanding of such patterns it is necessary to consider the patterns of competition in further detail. This can be achieved by looking initially at the **size** of the competition spaces, i.e. the number of schools competing with each other.

Size of Competition Spaces

Identifying the number of schools that compete against each other (i.e. those that gain from one another) provides an indicator for the size of the competition spaces. This information can be obtained from the matrix of gains and losses (Figure 4.1).

Table 4.2 shows the average number of schools that each school gains a significant number (i.e. more than 5) of pupils from (Column A). This gives an indication of the amount of competition going on. These patterns can be considered alongside the last column (C) in the table, which produces a ratio based on the average size of competition and the total number of schools within the authority. This gives an indication of the scale of competition against the largest competition space possible within a LEA. For example, if the ratio is equal to 1.00 then all the schools in the LEA gain from each other. If this was the case then it could be said that there is only **one** competition space in the Local Authority.

Table 4.2   Size of competition spaces, by LEA

LEA	Based on 'significant' gains*		
	A *Average size*	**B** *Total number of schools*	**C** *Average as a ratio (A/B)*
Inner London borough (East)	6.80	15	0.45
Large metropolitan borough (West Midlands)	6.21	61	0.10
Small metropolitan borough (West Midlands)	4.15	20	0.21
Metropolitan borough (Greater Manchester)	3.62	21	0.17
County (Eastern)	2.21	38	0.06
County (West Midlands)	3.28	29	0.11
All schools	*4.45*	*184*	*N/A*

*   Gains of five or more pupils.

From Table 4.2 it is clear that the average number of schools that any particular school competes with varies by LEA. This ranges from nearly 7 competing schools in the large metropolitan borough (West Midlands) to just over two competing schools, on average, in the Eastern county.

It might be assumed that the size of the competition spaces is dependent upon the total amount of market activity within the market place. However, the variety of typical competition space sizes for the Local Authorities does not necessarily reflect the amount of 'active' choice within each of those market places (see Table 4.3). For example, the metropolitan borough in Greater Manchester has a high proportion of 'non-locals' attending schools, indicating that a high number of parents are engaging with the market ('active' choice), but the average competition space is smaller than in the West Midlands' small metropolitan borough, which has a relatively small amount of 'active' choice taking place. This could suggest that the 'active' choice is concentrated in particular competition spaces or just between particular schools.

Table 4.3   'Active' choice, by LEA

LEA	Total Intake (1995/96)	Number of schools in study	% Not 'Local'*
Inner London borough (East)	2,202	15	72.66
Large metropolitan borough (West Midlands)	11,612	61	68.28
Small metropolitan borough (West Midlands)	3,386	20	47.73
Metropolitan borough (Greater Manchester)	3,665	21	60.52
County (Eastern)	6,988	38	28.69
County (West Midlands)	4,458	29	41.68

* Calculated by taking the proportion of pupils in a school's intake that live outside the thiessen generated local catchment area.

It is also clear that the average number of competing schools is not totally dependent on the total number of schools in the Local Authority. This can be seen in the varying ratios of competition sizes to the total number of potential schools by LEA (Table 4.2, Column C). If the extent of competition was totally dependent on just the number of schools available in the Local Authority then the ratio would generally remain the same in all of the Local Authorities. Consequently, it can be assumed that there are other factors preventing or perpetuating the amount of competition within the Local Authorities.

By examining the ratios (Table 4.2 , Column C) it appears that even though the large metropolitan borough in the West Midlands, on average, has relatively large competition spaces they are a long way from extending across the whole Local Authority. This can be compared with the Inner London borough (East), which has nearly all the schools competing with one another and producing a single competition space. In the case of the large metropolitan borough (West Midlands) this can be explained by the size of the Local Authority and reflects that a **threshold** of competition has been reached. In other words there is only a certain number of pupils each school can compete for and consequently it could be argued that, at least for the majority of schools, the competition spaces for state secondary schools are finite in size.

Following on from this general conclusion it can be seen that one of the factors affecting the amount of competition is linked to the spatial nature of the Authorities. For example, the more rural Local Authorities, county (Eastern) and county (West Midlands), have the lowest ratio of potential competition space sizes occurring. The exception to this would once more appear to be the urban large metropolitan borough in the West Midlands but, again, this is probably due to a threshold being reached.

The above Tables illustrate the average competition space size for each Local Authority. However, **within** each LEA the size of an individual school's competition space can vary a great deal (Taylor, 2000). These patterns of competition size tend to be related to the general characteristics of the schools with different competition space sizes.

Figures 4.3 and 4.4 illustrate the characteristics of schools according to the size of their competition spaces. The three important school characteristics considered are: GCSE examination performances; private-state scores;[2] and catchment sizes.

The trends for the average GCSE performance of schools by their size of competition space are of great interest (Figure 4.3). All the results indicate that if there is a relationship between competition space size and examination performance then it is a negative one, i.e. the larger the competition space the worse the examination performance. However, this conclusion is largely a product of aggregating all the schools together from different LEAs, since urban schools with larger competition spaces are more likely to have low examination performances because of the relatively disadvantaged social backgrounds of the children in these schools. Consequently, identifying differences in examination performance of schools according to their competition characteristics is only useful if comparing schools at a **local** scale.

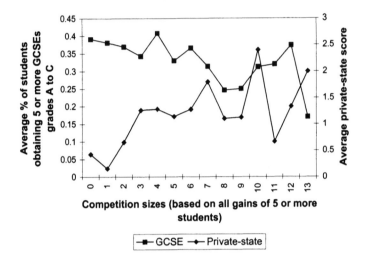

Figure 4.3 **Average GCSE performance and private-state scores by competition size of schools based on 'significant' gains, from a sample of eight LEAs**

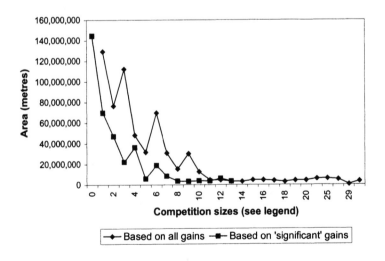

Figure 4.4 **Average catchment areas by competition size of schools, from a sample of eight LEAs**

These trends also suggest that the more 'private' schools appear to be either more competitive, more popular, or have very different competition spaces than 'state' schools (Figure 4.3). Because the results could indicate a number of things it may be necessary to separate out choices of schools based on the notion of a *better* school as opposed to the notion of *different* schools based on their type and diversity.

The third school characteristic considered in Figure 4.4, is the average catchment area size. It reveals the obvious downward trend in the school catchment size as they increase their extent of competition to more schools. Factors such as proximity and accessibility are key explanations for this and highlight the difference in urban and rural competition. However, on closer examination of the trend lines, the results are slightly erratic and do indicate that there are small catchment area schools with small competition spaces also.

This section has begun to outline the patterns of competition occurring across the different market places. The importance of urban-rural differences and the varying degrees of 'privateness' of schools has been highlighted as impacting upon the competition spaces. These results have also begun to suggest that there are different forms of competition spaces within the market place, but the patterns need to be further broken down into the different ways they might operate. This conceptualisation of the market place requires further analysis to consider which schools compete with one another and if there is some form of *organising principle* that determines which schools compete with one another.

The Form of Competition in the Market Place

Having identified some of the key characteristics that help to define the **competition spaces** it is necessary to highlight on what basis the schools compete with one another. This section first outlines three significant ways, or organising principles, in which schools can compete before going on to illustrate how this is achieved and to evaluate the importance of each form of competition.

In order to break down school-scale competition across the market place three sets of gains can be identified from studying the admission patterns of the schools in the study LEAs. Each of these provide a possible interpretation for competition between schools:

i.   The nearest alternative.
ii.   The examination league table.

iii.   The private-state continuum.

(I) Nearest school gains   The nearest school gains would suggest that most 'choices' occurring within a competition space are based on rejecting the local school in favour of the next nearest school. Under this force the competition spaces are likely to form discrete units within the market place (Figure 4.5), not dissimilar to the 'local competitive arena' (Woods *et al.*, 1996).

(II) Examination league table   Using the examination league table as a force behind competition it could be said that a school would only gain pupils from another school below them in the examination league table (Figure 4.6). This is a form of hierarchy but one based on the notion of a 'ladder of improvement'. In other words it is based on the parents' wish to improve the possibility of their child getting good examination results by choosing a school that has a set of pupils who generally do better than those in another school. Using the examination league table as a force for competition suggests that there is only a single hierarchy within the market place.

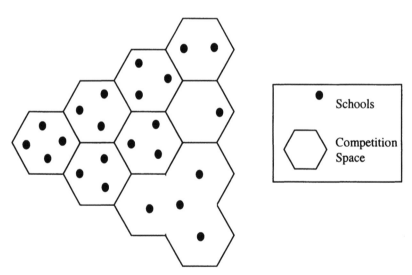

Figure 4.5 'Local' competition spaces based on nearest school gains

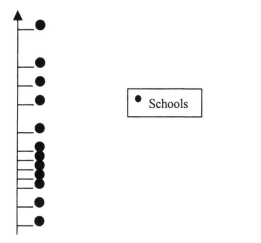

Figure 4.6   Examination league table and relative school positions

(III) Private-state continuum   The private-state continuum would suggest that the more 'private' schools, such as religious schools, gain from many other schools but actually only compete with schools of a similar type (Figure 4.7). This is based on diversity within the market place and that the competition spaces within the market place can be clearly labelled as pertaining to a particular type of education product. Consequently, parents and pupils choose the particular product they want and then make a preference from within all the possible schools that offer that product. These competition spaces exist in parallel to each other with little real competition between them since the pool of demand will be different for each competition space.

As Figure 4.7 illustrates, the spatial extent of each competition space can be the same, yet, within that same physical space there are different pools of customers. In terms of simple gains and losses between schools, as suggested earlier, these layered markets are hidden. Instead, it would appear that all schools compete with each other but the relative size of gains and losses between these layers would actually be quite small. As a result it should be acknowledged that the size of the gains and losses are very important when examining the competition patterns. In particular it would be necessary to differentiate between the competition patterns of net gains of **all** sizes and the competition patterns of net gains of **five or more** pupils (the 'significant gains') as discussed earlier.

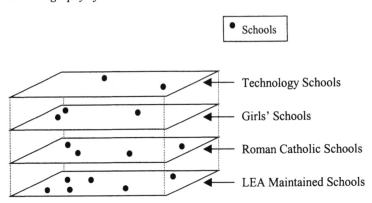

Figure 4.7 Parallel 'private' competition spaces based on diversity of choice

The patterns of competition can now be considered in terms of each of these concepts.

Nearest School Gains

The total number of gains occurring in the six Local Authority case studies was 17,719. Of these total gains, 4,170 were gains from the nearest alternative school. This suggests that 23.53%, nearly a quarter, of all gains took place between schools and their nearest competitor. (Note that this is not necessarily the nearest alternative school for the pupils.) These types of gains are about the schools and the competition between them and, therefore, focus on the exchanges between them. It is difficult to ascertain if this is a significant proportion of pupils or not. Remember that these *nearest school gains* are based on only *one* alternative school. This single school accounts for nearly a quarter of all gains to any school, and in these terms would appear to be significant. In relation to the model of the 'local competitive arena' this might reflect the expected amount of gains because the suggested *competition space* would incorporate approximately four adjacent schools.

There was a considerable degree of variation between Local Authorities, however, as can be seen in Table 4.4. The proportion of all gains as *nearest school gains* varied between 16% and 41%, in the small West Midlands metropolitan borough and Eastern county respectively. The Local Authorities with a high proportion of these gains were predominantly the more rural LEAs. For example, more than 40% of the competition in the

Table 4.4 Typology of competition as a proportion of all gains, by LEA

LEA	Total number of gains	Total gains (% of all pupils)	Nearest school gains (%)*	Up examination league table gains (%)*
Inner London borough (East)	1,600	72.66	24.94	54.56
Large metropolitan borough (West Midlands)	8,422	72.51	19.16	49.58
Small metropolitan borough (West Midlands)	1,616	47.73	16.58	56.87
Metropolitan borough (Greater Manchester)	2,218	60.52	20.87	74.71
County (Eastern)	2,005	28.68	41.85	68.28
County (West Midlands)	1,858	41.61	31.59	56.08
All schools	*17,719*	*54.84*	*23.53*	*56.64*

*   These are a percentage of the total number of gains.

Eastern county took place between adjacent schools. It is worth noting that these LEAs had a relatively small amount of competition occurring overall. Consequently, Table 4.5 illustrates these gains as a proportion of the total number of pupils within the Local Authority. From this table the conclusion is slightly different.

   As a proportion of all pupils, the Eastern county's *nearest school gains* were actually less than the *nearest school gains* in the urban large metropolitan borough (West Midlands), Inner London borough (Eastern) and small metropolitan borough (Greater Manchester). This might suggest that even if there was more active choice within the Eastern county then the increase in gains would not necessarily be *nearest school gains*.

   The expectation that the *nearest school gains* would provide a significant interpretation for competition in the context of rural schools is

Table 4.5   **Typology of competition as a proportion of all pupils, by LEA**

LEA	Total number of pupils	Total gains (% of all pupils)	Nearest school gains (%)*	Up examination league table gains (%)*
Inner London borough (East)	2,202	72.66	18.12	39.65
Large metropolitan borough (West Midlands)	11,615	72.51	13.90	35.95
Small metropolitan borough (West Midlands)	3,386	47.73	7.91	27.14
Metropolitan borough (Greater Manchester)	3,665	60.52	12.63	45.21
County (Eastern)	6,992	28.68	12.00	19.58
County (West Midlands)	4,465	41.61	13.15	23.34
All schools	*32,325*	*54.84*	*12.90*	*31.05*

*   These are a percentage of the total number of pupils.

further dispelled by looking at these kinds of gains by catchment type (Figure 4.8). This shows that it is the suburban schools that generally have the largest amount of competition with the nearest school. In these areas the proportion of gains from the nearest school was over 50%, that is one in every two 'gains' came from the nearest school.

Urban schools had a similar proportion of their gains coming from the nearest school to rural schools. (It should be noted that the fluctuations in the rural schools' results are possibly due to the small number of schools within those classifications.)

The reason why rural school gains did not feature predominantly in this form might be due to the key differences between accessibility and proximity. The choice of schooling in rural areas is likely to be constrained by travel to work routes, bus routes, etc. These differences are likely to be more important in rural areas than in urban areas and, therefore, might explain why competition between adjacent schools was less than schools that have greater accessibility to each other.

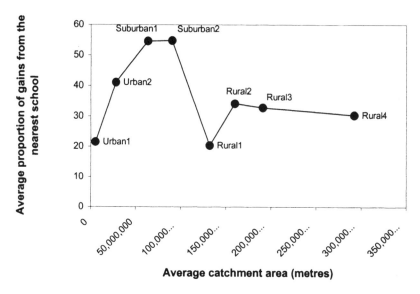

Figure 4.8 Nearest school gains by school catchment area types

A second explanation for this geographical distribution of nearest school gains might be due to the nature of the decision-making process. In an urban area there are a number of close (spatially) alternatives other than the nearest school, whilst in rural areas there is less available choice nearby and so the status quo is more likely. However, if an alternative school is desired in a rural area then the decision has to be considered carefully because of geographical constraints, which therefore reduces the inevitability of the final choice being the next nearest school. Suburban areas fall between the two. There is some choice of schools within close range, and, therefore, less significant constraint on the decision-making process than in rural areas, but less nearby choice than in urban areas. In suburban areas, therefore, it could be argued that the relative proximity of schools encourages activity in the market but are far enough apart to limit the alternatives. Consequently, this might explain why parents in suburban areas would often choose adjacent schools.

In the four urban LEAs none of the schools appeared to have a significant amount of competition with just their nearest school. There are some schools in these LEAs where approximately half of their gains came from the nearest school to them. But, for the majority of these cases the schools were located relatively very close to another school and, therefore, a good deal of competition took place between them. In some of these cases the level of competition between the neighbouring schools was largely one-way, where in others the competition between them operated in both directions.

However, altogether, five different cases were identified in which there was a high degree of competition between adjacent schools:

- school located between two other schools that are close by;
- school isolated on own or with one other school;
- two schools close to each other – one-way gains;
- two schools close to each other – two-way gains (usually single sex schools); and
- schools centrally located to many other schools – but is a poor performing school and the number of gains is very small.

There were schools in both urban and rural LEAs that had a high proportion of *nearest school gains* and were located near the edge of the Local Authority boundary, often in isolation. It is inevitable, therefore, that most of the gains for these schools will tend to come from the nearest school. It should be acknowledged that there may be some cross-LEA competition but this is generally at a minimum because of the bureaucracy associated with this.[3]

Examination League Table

As already suggested the *examination league table* could provide one way of interpreting the competition between schools. This creates a single hierarchy within the Local Authority and organises competition between schools based on this hierarchy. To identify such competition gains to schools higher in the examination league can be separated from gains that are to schools lower in the league. Table 4.4 showed that, for all schools in the study, 56% of gains were 'up' the examination league table. This is quite a significant proportion, particularly as it becomes harder for schools at the bottom of the league table to gain from schools with even lower examination results. This overall pattern was reproduced in each Local Authority because the figures ranged only between 45% and 75%. The metropolitan borough in Greater Manchester and the Eastern county stood out as having a very large proportion of gains up the examination league table.

The two Local Authorities with apparently the most 'active' choice, the Inner London borough (East) and the large metropolitan borough in the West Midlands, both had the lowest proportion of gains 'up' the respective LEA examination league table. This could indicate two things. One relates to the problem of having finite supply and demand in the education market places. As there is more demand for schools and greater active choice then the supply of surplus places in those schools falls. Consequently, the only surplus places in the system allowing greater choice for parents are in schools lower down in the examination league table. The second interpretation of such patterns is that the parents and children in these Local Authorities do not consider the examination league table as an important way of providing an organising principle for their choice of schools, i.e. other factors are used. It is also worth noting that the Inner London borough (East) and the large metropolitan borough in the West Midlands had the lowest average GCSE examination performances of the study LEAs, and would therefore normally suggest that choosing a school with a good examination performance would be important.

These interpretations can be considered alongside the number of gains up the examination league as a proportion of all pupils within each Local Authority. According to Table 4.5 the Inner London borough (East) and the large metropolitan borough in the West Midlands are among the LEAs with the highest proportion of such gains. This suggests that the first interpretation of finite supply and demand, given above, might be more prevalent, because, in absolute terms, these two LEAs did have more pupils going to schools with better results than, say, in the two County LEAs.

Even with these absolute figures, the Greater Manchester metropolitan borough remained the Local Authority with most competition being associated with improvement up the examination league table. This could be due to the attitude and decisions of the parents and children as to what is important in school choice.

Examining these gains by catchment type supports what has already been hinted at, that the patterns of an examination league table hierarchy are not significantly different in urban or rural contexts (Figure 4.9). (The fluctuations in the rural school results are probably due to the small number of schools in each category.) It is interesting that the significance of the examination league table remained fairly constant between the LEAs even though access and proximity varies between the Local Authorities. This can only further enhance the importance of the examination league table in understanding competition in the education market overall.

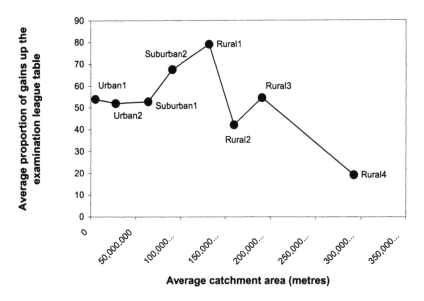

Figure 4.9 Gains up the examination league table by school catchment area type

'Private' Gains

Of all the gains and competition that took place, 41% were gains to what could be considered as the more 'private' schools (i.e. their private-state score > 0), and the variations by Local Authority were quite dramatic (Table 4.6). For example, in the Inner London borough (East) over 60% of all gains were to 'private' schools, but in the Eastern county only 12% were to these more 'private' schools. However, the unequal number of 'private' schools within each Local Authority seriously distorts these figures. Therefore, these *'private' gains* need to be considered alongside the proportion of 'private' schools within the Local Authority.

In this study 35.87% of all schools had some degree of 'privateness' (i.e. their private-state score was greater than zero). These schools in turn had 40.89% of all the gains, or competition. The relatively small differences between these figures suggest that competition did not exclusively, nor significantly, belong to 'private' schools. However, they did have a small number of gains above their *expected* share.

The difference between the *actual* and the *expected* proportion of 'private gains' by Local Authority is quite informative. If it were assumed that competition is distributed equally among schools then there would be no difference between the expected and actual proportion of *'private'* gains. Table 4.6 shows that in the large metropolitan borough (West Midlands) there were fewer *private gains* than perhaps expected. This differs considerably from the West Midlands county and the metropolitan borough in Greater Manchester. In these two Local Authorities the presence of *'private'* gains was 9% and 14%, respectively, above the expected occurrence. This suggests that the importance of 'private' schools in these two LEAs bear a great influence upon competition.

Table 4.6   'Private' gains, by LEA

LEA	'Private' gains (%)[a]	'Private' schools (%)[b]	'Expected' – 'Actual'	'Private' gains (%)[c]
Inner London borough (East)	60.06	60.00	0.06	43.64
Large metropolitan borough (West Midlands)	38.86	45.90	-7.04	28.18
Small metropolitan borough (West Midlands)	46.97	45.00	1.97	27.44
Metropolitan borough (Greater Manchester)	52.12	38.10	14.02	31.54
County (Eastern)	12.17	10.53	1.64	3.49
County (West Midlands)	36.71	27.59	9.12	15.27
All schools	*40.89*	*35.87*	*5.02*	*22.41*

a   These are a percentage of the total number of gains
b   These are a percentage of the total number of schools
c   These are a percentage of the total number of pupils

Figure 4.10 shows the average number of gains to the more 'private' schools against the average number of gains for all schools by catchment size. The pattern of *'private' gains* by catchment size follows the trend for all school gains, and hence it appears that there was no difference in the importance of 'private' competition across the urban-rural divide.

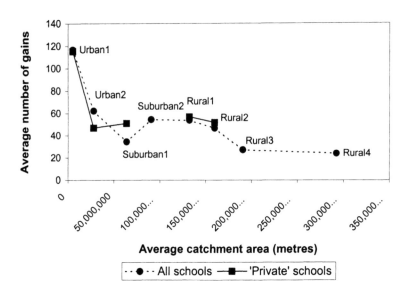

Figure 4.10 'Private' gains by school catchment area size

Categorisation of Competition Spaces

Having considered these three important types of competition it is possible to begin to construct a typology for the *competition spaces*. Based on the understanding of school competition patterns outlined above, and studying the intake and competition patterns of every school in six LEAs, it was possible to identify which schools competed with one another. (Examples of such competition spaces are given in Figures 5.1 to 5.4, Chapter 5.) The resulting competition spaces could then be defined and categorised according to the way in which competition was being organised. This highlighted three forms of competition space; private, hierarchical and non-hierarchical competition spaces. It was then possible to classify each school's position within these three forms of competition spaces according to how well they competed with other schools in their competition space. The resulting typology of competition spaces is given below:

i.   'Private' competition spaces – diversity in the market place
  - A - Competition extends across more than half of the LEA territory
  - B - Competition is significantly spread but less than half of the LEA territory
  - C - Competition is relatively local in nature
    - Ca - Cannot compete with better[a] state schools
    - Cb - Can compete with better[a] state schools
  - D - Grant Maintained schools competing in a 'state' hierarchy

ii. Hierarchical competition spaces
  - Ea - Top of a hierarchy
  - Eb - Top of a small hierarchy [b]
  - Fa - Middle of a hierarchy
  - Fb - Lower middle of a hierarchy
  - Ga - Bottom of a hierarchy
  - Gb - Bottom of a small hierarchy [b]

iii. Non-hierarchical competition spaces
  - H - Equal competition between schools
  - J - No competition between schools of the same 'type' (i.e. parallel competition only)

a   Based on the GCSE examination performance of the schools.
b   A hierarchy made up of only two schools or with a relatively small amount of competition between schools.

The three different forms of competition spaces, and the characteristics of schools that belong in them, are now discussed in turn before concluding with the different characteristics of competition that arise in these competition spaces.

'Private' Competition Spaces

'Private' competition spaces represent diversity within the market place and closely resemble the conceptual and observed patterns of competition outlined earlier. The spatial extent of competition for these schools varied from being totally LEA-wide to operating at a relatively local scale. This is reflected in their competition size, as seen in Figures 4.6 and 4.7. Consequently, 'private' schools do not just exist in a single parallel market to state schools. Instead it was possible to typically identify three layers of competition, each having a different degree of importance placed on the locale of the school.

The three 'private' competition spaces were only identifiable by their actual competition size and not by any identifiable **type** of 'private' schooling that they offered, such as single sex, Roman Catholic, or selective school types. However, as Figure 4.7 showed, there was a positive relationship between the spatial extent of competition and the private-state score of the schools. So, for example, the more 'private' a school was the larger the competition space and consequently the less importance the locale of the school had in this form of competition space.

Another distinction between the three 'private' competition spaces was in their GCSE examination performances. Table 4.7 shows that the greater the areal extent of the 'private' competition spaces the greater the GCSE examination performance of the schools.

Table 4.7   **Average GCSE performance by the spatial extent of the 'private' competition spaces**

LEA	Average proportion of pupils obtaining 5+ GCSEs grades A-C			
	All 'Private' schools	*A*	*B*	*C*
Inner London borough (East)	33%	44%	30%	20%
Large metropolitan borough (West Midlands)	32%	80%	33%	25%
Small metropolitan borough (West Midlands)	52%	70%	34%	N/A
Metropolitan borough (Greater Manchester)	47%	58%	56%	37%
County (Eastern)	49%	N/A	54%	46%
County (West Midlands)	62%	85%	60%	29%

A   Very large extent (competition space A)
B   Large extent (competition space B)
C   Local extent (competition space C)

The fourth type of 'private' competition space identified specifically involves GM schools that compete locally and even appear to compete within a hierarchy of 'state' schools (see below). The greater tendency for these 'private' schools to compete alongside 'state' schools probably reflects the recent addition of GM schools on to the education landscape. This suggests that hierarchical competition spaces have developed over many years, which nearly all GM schools were once part of. As a result, it is possible that some 'private' schools, GM schools in particular, compete locally enough to be considered as being at the top of a hierarchical competition space. It is clear, however, that GM status has provided an intermediary position between 'state' schools and the traditional 'private' schools, such as religious schools or single-sex schools.

Hierarchical Competition Spaces

Competition between some schools formed, what appeared to be, structured networks of gains and losses. These structured networks closely resembled a hierarchical organisation of competition, and tended to be distinct from one another. However, the number of schools in a hierarchy, the number of tiers in a hierarchy and the complexity of these hierarchies did vary between and within Local Authorities.

There were generally two common kinds of hierarchy identified. The first was a multi-tiered hierarchy in which schools gained from all other schools below them in the hierarchy. Usually the number of tiers in this kind of hierarchy did not exceed three, but there were some four-tiered hierarchies. The second kind of hierarchy was single-tiered, involving just two schools, but which was sometimes connected to another single-tiered hierarchy.

The distribution of these kinds of hierarchies varied between Local Authorities. In nearly all the rural and remote areas the single-tiered hierarchy was prominent, but in urban areas both the single-tiered hierarchies and the multi-tiered hierarchies could be found. Urban school hierarchies also tended to be more complex because the individual hierarchical competition spaces were nearly always connected to other hierarchical competition spaces in some way. One particular hierarchy in the West Midlands large metropolitan borough illustrated this very well by having 21 schools in its membership. However, this very large single connected competition space never exceeded four tiers and could be considered as having six connected individual hierarchies. This complex picture of competition in the large metropolitan borough (West Midlands) contrasts markedly with competition in the Inner London borough (East).

In this latter LEA about 60% of the schools used in this study were 'private' and consequently there were not many 'state' schools to form hierarchies, and so were separated out into small discrete hierarchies by the locations of the 'private' schools.

The majority of schools in hierarchical competition spaces were more 'state' schools. As mentioned above this does not mean that there were no 'private' schools competing locally for the same pupils as these 'state' schools in a hierarchy. Nor does this exclude the few 'private' schools, usually GM schools, which were occasionally at the top of these hierarchical competition spaces.

The position of the more 'state' schools in the hierarchies, i.e. which tier they were on, was largely based on their individual GCSE examination performance. Table 4.8 shows the average examination performance of schools depending upon whether they were at the top, middle or bottom of the hierarchical competition spaces. Of the 98 schools identified as competing in hierarchies, of some form, only 21 schools had positions in the hierarchies of competition that did not reflect their GCSE examination performance relative to their competitors. This is only just over one in five schools in a hierarchy. However, this did vary between Local Authorities, for example the large West Midlands metropolitan borough had a relatively high occurrence of these 'inconsistencies', but the small West Midlands Metropolitan borough had none at all. Interestingly, Tables 4.5 and 4.6 also highlighted the relative low importance of GCSE examination performance as an explanation for gains in the large West Midlands metropolitan borough.

The competition between schools within a hierarchy was usually one-way, i.e. a school rarely lost pupils to a school below it in the hierarchy. However, there were some instances of significant two-way gains and losses between schools within the same hierarchical competition space. Of the 98 schools identified as competing in hierarchies, 9 pairs of schools (18 schools in total) had a significant number of gains *and* losses between them, irrespective of their position in the hierarchies. This 'exchange' of pupils does not necessarily alter the structure of the hierarchy because in terms of net gains competition was still in the 'right' direction. It does show, however, that on the one hand these hierarchies can be fairly complex, but on the other it does mean that the identification of hierarchies for four in every five schools was done with some confidence.

Table 4.8 **Average GCSE performance by position in hierarchical competition spaces, by LEA**

LEA	Average proportion of pupils obtaining 5+ GCSEs grades A-C			
	All 'state' schools	*E*	*F*	*G*
Inner London borough (East)	22%	32%	N/A	12%
Large metropolitan borough (West Midlands)	19%	26%	17%	16%
Small metropolitan borough (West Midlands)	20%	36%	19%	13%
Metropolitan borough (Greater Manchester)	35%	50%	33%	27%
County (Eastern)	44%	50%	46%	38%
County (West Midlands)	37%	46%	27%	28%

E Top of hierarchy (competition spaces E and Ea)
F Middle of hierarchy (competition spaces F and Fa)
G Bottom of hierarchy (competition spaces Ga and Gb)

Non-Hierarchical Competition Spaces

Within this study, 98 'state' schools (82% of all 'state' schools) were identified as competing within some form of hierarchical competition space. This compares with 22 'state' schools (18% of all 'state' schools) that could not be identified as operating within a hierarchical competition space. These 22 schools could be split in to two groups, those that competed equally with another 'state' school (competition space H), and those that had no competition with other 'state' schools at all (competition space J). Eight schools, or four pairs of schools, were identified as only competing with each other and with an equal exchange of pupils between them. Consequently, the net result of their competition maintains the status quo in terms of pupil numbers. Fourteen schools in total were identified as not being in any competition with other 'state' schools at all. However, it should be noted that schools in the parallel 'private' competition spaces did impact upon the intakes of **all** these schools. The majority of such schools were in the two more rural counties, but there were three from urban areas.

However, whether they were in urban or rural locations it was the schools' relative isolation that probably prevented any competition taking place between them and other 'state' schools.

Conclusion

This Chapter has tried to provide a means of interpreting the education market place using admissions data from a sample of LEAs. In doing so it has attempted to reduce the complexity of the market so that it is possible to understand some of the main ways in which the market place operates, i.e. what competition there is between schools and what organising principles can be used to conceptualise competition in the education market. However, using the notion of competition spaces, the context and the geographical specifics of the many market places can be considered.

It has been shown that the market place needs breaking down into competition spaces so that real occurrences of competition, 'gains' and 'losses' of pupils between schools, can be identified. Several constraints on competition were noted, in particular, proximity to other schools, the urban-rural context to the market place, and the accessibility of schools to parents. Patterns of competition were then illustrated indicating the complexity of the market place before comparing the characteristics of schools against the extent, or size, of competition they faced. Using a number of techniques for measuring and observing competition between schools a number of conclusions were noted:

- the typical number of schools that compete with each other is between 9 (using all gains) and 4 (using 'significant gains');
- the levels of market activity and size of LEA are not the only factors determining the size of *competition spaces*;
- rural LEAs tend to have smaller *competition spaces* than urban LEAs;
- there might be a limit to the extent of competition between schools, particularly in urban LEAs; and
- 'private' schools tend to compete with more schools than 'state' schools, but for fewer pupils.

Based on this discussion on the patterns of competition spaces the analysis then focussed on the processes of competition between schools. Three key organising principles of competition were determined from the data:

i.   Competition based on the nearest alternative school.

ii.   Competition based on the examination league table.
iii. Competition based on the presence of 'private' schools within the market place.

These three forms of competition were examined to determine their relative importance in the market place. All were deemed significant but they also illustrated the differences between Local Authorities. The nearest school gains varied between urban, suburban and rural schools illustrating the problems of access and proximity. The examination league gains began to show that as the market develops and more parents are active in the market place there was a limit to schools competing with one another based on this principle. And 'private' gains varied between Local Authorities irrespective of their geography possibly reflecting the differing impact of the private-state continuum on the market place.

This analysis then led to providing a categorisation of all schools based on three identifiable forms of *competition spaces*:

i.   Private competition spaces.
ii.   Hierarchical competition spaces.
iii. Non-hierarchical competition spaces.

Most importantly, this showed the various ways in which schools actually compete with one another. For example, private schools tend to compete in *parallel* markets to each other, closely related to the degree of their 'privateness' and their examination performances. It was also shown that for the majority of 'state' schools they compete with each other in localised hierarchies that are closely related to the schools' relative examination performances. The Chapter concluded with a presentation of some important, geographical, characteristics that impacted upon schools in all three forms of competition spaces.

Finally, this discussion has provided the necessary insight into how the new education market operates, and the context in which to examine the choices and behaviour of parents (Chapter 6) and the impact on social inequality and inequity due to the reforms in education (Chapter 7). First, however, the next Chapter presents the competition and socio-economic characteristics of intakes of eight case study schools from the study.

Notes

1   The LEAs are described as an Inner London borough (Eastern), an Inner London borough (Western), an Outer London borough (Northern), a large metropolitan borough

(West Midlands), a small metropolitan borough (West Midlands), a metropolitan borough (Greater Manchester), a county (Eastern), and a county (West Midlands). The admissions data for two of the London boroughs, the Outer London borough (Northern) and the Inner London borough (Western), were rather incomplete because a significant number of schools did not provide admissions data directly to the LEAs. Therefore on occasions in this discussion these two Authorities were not used in the analysis.

2   Private-state scores were outlined and discussed in Chapter 3.

3   See the beginning of Chapter 3 for a discussion on cross-LEA competition.

5 Competition and Choice: Eight Case Studies

Introduction

The previous Chapter identified how schools across eight different LEAs appeared to be competing against one another. This Chapter begins to examine the market place at a more detailed level by focussing on eight schools from three different LEAs, each representing different features of the 'new' education market place outlined in the previous Chapter. The aims of this Chapter are to present the characteristics of these study schools in three stages. First, it will illustrate the educational and competition characteristics of the schools using published data and the analysis of competition from the previous Chapter. The Chapter will then present the socio-economic characteristics of the intakes of these eight schools based on the postcode admissions data. The third stage will discuss the socio-economic characteristics of different sets of pupils that comprise the schools' intakes in order to make some preliminary observations of changes to the intakes of these schools due to the policy of open enrolment.

Overall, therefore, this Chapter will outline the characteristics and intakes of the eight schools in order to provide the context for the following Chapter, which focuses on the household decision-making process for choosing a secondary school.

The eight schools that form the basis of the analysis in this Chapter are:

Albert Randall School (ABRA)	Metropolitan Borough (Greater Manchester)
Pembroke Community College (PEMB)	Metropolitan Borough (Greater Manchester)
St Percival's Roman Catholic School (STPR)	Metropolitan Borough (Greater Manchester)
Hengrove School (HEN)	County (West Midlands)

Polden School (POL)   County (West Midlands)

Hartland School (HAR)   County (Eastern)

Northleigh School (NOR)   County (Eastern)

Thorndale School (THO)   County (Eastern)

School Characteristics

The characteristics of the schools considered here were divided into two components. The first involved their educational characteristics including GCSE examination performance, the proportion of the pupils in each school who had been statemented for special needs teaching, the level of authorised absences across the whole school, and the degree to which the schools were controlled and influenced by the 'private' sphere, as discussed in Chapter 3. Apart from the private-state score, which was calculated especially for this study, the other educational characteristics are published in annual school performance tables. The second set of school characteristics considered in this Chapter included indicators, or competition characteristics, of each schools' performance in the market place. This was based upon the typology of competition identified and defined in the previous Chapter. The precise nature of each school's particular **competition space**, such as which schools they appeared to compete with and how that competition was organised, is presented later in the discussion. However, these forms of competition space are presented alongside two further indicators of competition that identified each school's ability to, both, *attract* pupils from outside their own 'local' catchment area, and to *successfully* prevent the 'local' catchment pool of pupils from attending an alternative school. These competition characteristics are also presented alongside the spatial nature of each school's catchment area based on spatial analysis presented in Table 4.1 (Chapter 4).

These characteristics are summarised in Tables 5.1 to 5.6 (by LEA) as well as the respective LEA averages for each characteristic. The fact that these eight schools were from three different Local Authorities provides a basis for comparison between different market places in much greater depth. However, the schools represent different examples of the relationship between their characteristics, their market performance and the composition of their intakes. It is this latter context, within each Local Authority, that the discussion now concentrates on.

'Parallel' Competition in a Greater Manchester Metropolitan Borough LEA

The three schools from a LEA in Greater Manchester used in this part of the research, St Percival's Roman Catholic School, Albert Randall School and Pembroke Community College, all represent urban schools and were all in 'direct' or 'indirect' competition with each other. This was due to their close proximity to one another. However, these three schools illustrated the effects of competition across, and between, 'private' and 'state' competition spaces, since St Percival's was in parallel competition to the other two schools. On the one hand, this school attracted pupils from a very large area of the LEA, well beyond the catchments of Albert Randall and Pembroke, but, on the other, it still depended on attracting pupils from the catchment areas of the other two 'state' schools. In turn, Albert Randall and Pembroke were in the same 'local' state competition space (Figure 5.1), in which Albert Randall was at the top of the 'state' hierarchy and Pembroke was at the bottom (Table 5.1).

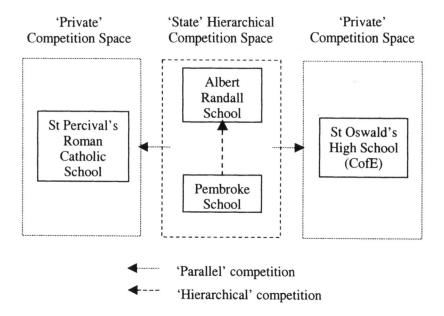

| 'Private' Competition Space | 'State' Hierarchical Competition Space | 'Private' Competition Space |

◄┄┄┄┄ 'Parallel' competition

◄╶╶╶ 'Hierarchical' competition

Figure 5.1 Competition spaces of Albert Randall, Pembroke and St Percival's schools, Greater Manchester metropolitan borough

Table 5.1   **Competition characteristics of three schools in a Greater Manchester metropolitan borough LEA, 1995**

School	Area type[a]	Competition Space[b]	Attraction[c]	Success[d]
LEA Average	-	-	*56.2%*	*38.8%*
St Percival's (STPE)	Urban	A Private – half of LEA	92.9%	17.0%
Albert Randall (ABRA)	Urban	Ea Top of a 'state' hierarchy	89.7%	29.4%
Pembroke (PEMB)	Urban	Ga Bottom of a 'state' hierarchy	28.6%	26.0%

a   As defined in Table 4.1 (Chapter 4)
b   As defined in Chapter 4
c   Proportion of intake from outside the 'local' 'thiessen' catchment area (see Chapter 4)
d   Proportion of 'local' 'thiessen' catchment area intake that continued to attend their 'local' school

All three schools' position in their respective competition spaces was related to their differing levels of *attraction* and *success* (Table 5.1). St Percival's was very attractive to pupils outside its immediate locale, but the identification of this school as in a parallel 'private' competition space clearly explains the very low success of this school in keeping pupils from within its immediate locale. Similarly, Albert Randall was nearly as attractive outside its 'local' catchment area and was far more successful in keeping its 'local' intake. However, the impact of surrounding 'private' schools, such as St Percival's, keeps this success score below the LEA average. The poor market performance of Pembroke was observable from its *attraction* and *success* indicators; it was neither very attractive to pupils outside its 'local' catchment nor successful at keeping its 'local' catchment.

   The market performance of these schools was reflected in their respective GCSE examination performances, in which St Percival's had the greatest proportion of pupils obtaining 5 or more GCSEs with grades A to C, followed by Albert Randall and then Pembroke (Table 5.2). The position of Albert Randall in its competition space illustrated the above average performance of this school in GCSE examinations. The other two educational characteristics also reflected the respective 'advantages' of each school in the market place, such that Pembroke had the highest levels of special needs and the greatest level of absences to go alongside with its low market and examination performance.

Table 5.2   Educational characteristics of three schools in a Greater Manchester metropolitan borough LEA, 1995

School	GCSE[a]	Statemented[b]	Absences[c]	Private-State score[d]
LEA Average	*42%*	*2.3%*	*8.7%*	*0.06*
St Percival's (STPE)	62%	1.9%	8.5%	2
Albert Randall (ABRA)	46%	3.7%	9.6%	0
Pembroke (PEMB)	13%	3.9%	11.9%	0

a   Proportion of 15 year olds with 5 or more GCSEs with grades A to C
b   Proportion of pupils in school statemented for special needs teaching
c   Number of authorised half-day absences as a percentage of number of school roll
d   As defined in Chapter 3

The three schools in this Greater Manchester metropolitan borough provide an example of, on the one hand, schools competing *across* different types of competition spaces and, on the other, schools competing *within* the same competition space.

Limited Competition in a West Midlands County LEA

Both the West Midlands county schools, Hengrove School and Polden School, represent schools that appeared to be in no 'direct' competition with other 'state' schools (Table 5.3). This did not mean that they were not in any indirect competition with 'private' schools, but if they were then this would not have been of significant magnitude alone to affect the size and composition of their respective intakes. Even though they were in no competition with other 'state' schools they were from two different areas of the LEA and were classified in different types of area; Hengrove was in a rural location, whereas Polden was located in a more suburban area of the county.

Table 5.3 Competition characteristics of two schools in a West Midlands county LEA, 1995

School	Area type[a]	Competition Space[b]	Attraction[c]	Success[d]
LEA Average	-	-	*46.6%*	*53.0%*
Hengrove (HEN)	Rural	J No Competition	29.8%	69.7%
Polden (POL)	Suburban	J No Competition	1.1%	90.3%

a   As defined in Table 4.1 (Chapter 4)
b   As defined in Chapter 4
c   Proportion of intake from outside the 'local' 'thiessen' catchment area (see Chapter 4)
d   Proportion of 'local' 'thiessen' catchment area intake that continued to attend their 'local' school

From the educational characteristics of these schools it is clear that Hengrove was a school that could have been more attractive since it had an above average GCSE examination performance and a low proportion of pupil absences (Table 5.4). Similarly, Hengrove's *attraction* indicator was significantly higher than Polden's, possibly reflecting its greater potential market performance. Clearly, the isolated location of this school was limiting the number pupils who could have chosen it. However, even though Polden was in a suburban area, i.e. closer to neighbouring schools than Hengrove, the levels of *attraction* and *success* reflected much higher market isolation. Consequently, these two schools represented both, examples of the relationship between the geography of the market place and their potential performance in the market place within the same LEA, and examples of schools with limited competition between schools and, consequently, limited choice for parents.

Table 5.4   **Educational characteristics of two schools in a West Midlands county LEA, 1995**

School	GCSE[a]	Statemented[b]	Absences[c]	Private-State score[d]
LEA Average	*43%*	*1.0%*	*7.4%*	*0.32*
Hengrove (HEN)	48%	1.9%	5.7%	0
Polden (POL)	37%	1.0%	8.7%	0

a   Proportion of 15 year olds with 5 or more GCSEs with grades A to C
b   Proportion of pupils in school statemented for special needs teaching
c   Number of authorised half-day absences as a percentage of number of school roll
d   As defined in Chapter 3

The Urban-Rural Contrast in an Eastern County LEA

The three schools, Northleigh School, Thorndale School and Hartland School, provided a basis for comparing schools from within the same authority but were located in different geographical areas. Northleigh was an urban school in the county's principal city, whereas Thorndale and Hartland are located in the countryside (Table 5.5). There were also geographical differences in the location of these two rural schools; Thorndale was located in a relatively large market town and was, therefore, much more accessible for pupils than Hartland. All three schools had a very similar proportion of pupils obtaining 5 or more GCSEs with grades A to C (Table 5.6), yet performed differently in the market place. For example, even though Northleigh was an urban school its ability to attract pupils from beyond its 'local' catchment area was similar to that of Thorndale. In turn, Thorndale was a great deal more attractive than Hartland, the other rural school. The different locations of these three schools also reflected differences in their competition spaces. (Figures 5.2 to 5.4). Their respective degrees of proximity to, and isolation from, other schools provided a real comparison of the variations between urban and rural schools within the same market place. Since they were all relatively popular schools, with similar educational characteristics, variations in their relationship with the market place is much more likely to be due to the geographical constraints upon competition and choice.

Table 5.5 **Competition characteristics of three schools in an Eastern county LEA, 1995**

School	Area type[a]	Competition Space[b]	Attraction[c]	Success[d]
LEA Average	-	-	*29.9%*	*71.6%*
Northleigh (NOR)	Urban	Ea Top of a 'state' hierarchy	51.2%	79.0%
Thorndale (THO)	Rural (market town)	Ea Top of a 'state' hierarchy	52.1%	80.7%
Hartland (HAR)	Rural	Fb Middle of a 'state' hierarchy	13.0%	71.9%

a   As defined in Table 4.1 (Chapter 4)
b   As defined in Chapter 4
c   Proportion of intake from outside the 'local' 'thiessen' catchment area (see Chapter 4)
d   Proportion of 'local' 'thiessen' catchment area intake that continued to attend their 'local' school

Table 5.6 **Educational characteristics of three schools in an Eastern county LEA, 1995**

School	GCSE[a]	Statemented[b]	Absences[c]	Private-State score[d]
LEA Average	*44%*	*2.6%*	*7.3%*	*0.25*
Northleigh (NOR)	57%	1.0%	7.9%	0
Thorndale (THO)	58%	1.7%	6.1%	0
Hartland (HAR)	56%	1.7%	5.4%	0

a   Proportion of 15 year olds with 5 or more GCSEs with grades A to C
b   Proportion of pupils in school statemented for special needs teaching
c   Number of authorised half-day absences as a percentage of number of school roll
d   As defined in Chapter 4.4

'State' Hierarchical Competition Space

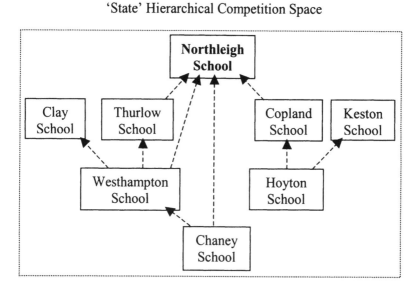

<----- 'Hierarchical' competition

Figure 5.2   Competition space of Northleigh School, county LEA (Eastern)

'State' Hierarchical Competition Space

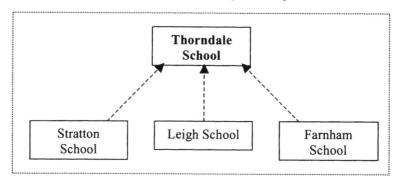

<----- 'Hierarchical' competition

Figure 5.3   Competition space of Thorndale School, county LEA

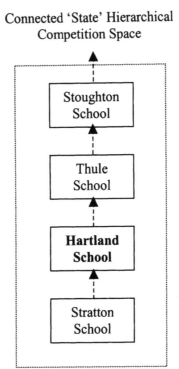

Connected 'State' Hierarchical
Competition Space

◄--- 'Hierarchical' competition

Figure 5.4 Competition space of Hartland School, county LEA (Eastern)

Intake Characteristics

Before analysing the in-depth responses of individual parents in the next Chapter it is useful at this stage to examine the socio-economic characteristics of each of the eight case study schools' intakes. Using the geodemographics package, GB Profiler '91, the home postcode of each pupil was used to identify a relative measure of eleven socio-economic variables generated from a 100-fold classification scheme based on the 1991 Census (see Taylor, 2000). The GB Profiler '91 variables used in this analysis were as follows:

- young people;
- older people;
- ethnicity;
- privately owned households;
- rented households;
- white-collar workers;
- blue-collar workers;
- young families and couples;
- older families and couples;
- wealth indicators; and
- poverty indicators.

Each of these variables was indexed such that the UK average is equal to 100. As a result each pupil that attended a particular school was assigned a value for each of these variables based on where they lived. These values were then aggregated to produce a socio-economic profile of each school's intake.

The socio-economic characteristics of all pupils in the three LEAs tended to reflect general differences between urban and rural areas of England. The households of pupils from both the county LEAs had above UK average levels of wealth Indicators and white-collar workers. The 1995 intake of pupils the Greater Manchester metropolitan borough came from less prosperous and less professional neighbourhoods. All three authorities had pupils with similar levels of ethnicity, relatively low for the UK, and lived in similar forms of accommodation. The intakes from the Eastern county tended to come from areas with high blue-collar workers, reflecting the higher levels of agricultural employment in the country.

Within each Local Authority the variations between the socio-economic composition of the schools in the sample were much greater. For example, there were great differences in the accommodation, employment, age and wealth of the intakes of the three schools in the Greater Manchester LEA, and was related to each school's market performances. The intake of St Percival's was from neighbourhoods with the highest level of privately owned accommodation, the greatest level of white-collar workers, had more 'mature' households, and had the highest levels of wealth indicators. This contrasted quite dramatically with Pembroke, the least attractive of the three Greater Manchester schools, and from the bottom of a 'state' hierarchy. The pupils attending Pembroke were generally from neighbourhoods with the lowest levels of privately-owned accommodation and the highest levels of rented accommodation. They also had low white-collar workers *and* blue-collar workers index scores suggesting that they lived in areas with high unemployment. This was also alongside younger families and couples and high levels of poverty. The significance of the variations in the intakes of the three Greater Manchester schools was further enhanced by the fact that they are all in the *same* area and, all things being equal, would theoretically compete for the *same* pupils.

The socio-economic composition of the two West Midlands county schools, Hengrove and Polden had less disparity, but Hengrove tended to attract pupils with relatively more socially 'advantaged' home backgrounds. This was seen in the higher levels of privately owned accommodation, higher presence of white-collar workers, older, more mature, families, and higher levels of wealth for neighbourhoods of where pupils attending Hengrove came from. Obviously, these distinctions between the intakes of Hengrove and Polden might have simply reflected the unequal distribution of social 'advantage' and 'disadvantage' of pupils across the county.

In the previous section it was argued that the schools in the Eastern county provided examples of different geographical contexts to competition and choice. Certainly the socio-economic characteristics of the pupils who attended Hartland reflected the rural and isolated location of this school. The pupils at this school tended to be from neighbourhoods with relatively low levels of ethnic diversity and high levels of blue-collar workers, reflecting the presence of agricultural employment in the community. They also tended to come from more mature families and had high levels of wealth in comparison to pupils attending other schools. However, perhaps rather surprisingly, the pupils attending the urban school of Northleigh tended to be from more privileged backgrounds than those pupils who attended Thorndale, the rural market town school. Pupils at Northleigh School tended to come from neighbourhoods with higher levels of privately owned accommodation, a higher presence of white-collar workers and higher indicators of wealth than those from Thorndale.

Again, as in the West Midlands county, it was not necessarily clear if these differences in the composition of intakes simply reflected socio-economic differences in the different parts of the Eastern county. In order to account for these variations, and to begin to understand how the intakes are being composed of pupils from different socio-economic backgrounds, the next section focuses on the characteristics of different sets of pupils that make up the whole intake of each school.

Intake Profiles

To begin an analysis of how intakes are formed through the process of choice the pupils were divided into particular subsets, or profiles, of the schools' intakes. For each school it was recognised that there were five different **intake profiles** that could be used to identify the changes to the socio-economic composition of the intakes. They were all based on a pupil being classed as either, 'local' to the school they attended, i.e. they lived in the 'thiessen' catchment area of the school, or as a 'non-local', i.e. they lived outside the school's 'thiessen' catchment area. Based on this supposition and definition of a 'local' and a 'non-local' the following five **intake profiles** were identified as:

Actual intake	– the *actual* intake of school X
Catchment intake	– only the *local* students who went to school X
Outside intake	– only the *non-local* students who went to school X
Lost intake	– *locals* of school X but who *did not* go to school X
All catchment intake	– All the *local* students of school X, including those that did not go to school X

These intake profiles are illustrated in Figure 5.5, which provides a hypothetical example of School X and its 'thiessen' catchment area. The first diagram shows the intake of School X (the 'actual' intake profile) and the pupils who lived in the catchment area but who attended an alternative school. The second diagram illustrates how three intake profiles are defined from the same intake distribution for School X. From these three intake profiles the other two intake profiles, 'actual' and 'all' catchment intake profiles can be defined using the following two formulas:

'Actual' intake = 'Catchment' intake + 'Outside' intake
'All' catchment intake = 'Catchment' intake + 'Lost' intake

From these intake profiles there were three different comparisons that could be made to summarise the changes to the socio-economic composition of each school's intake:

i.   'Catchment' intake vs 'lost' intake.
ii.   'Outside' intake vs 'catchment' intake.
iii. 'Outside' intake vs 'lost' intake.

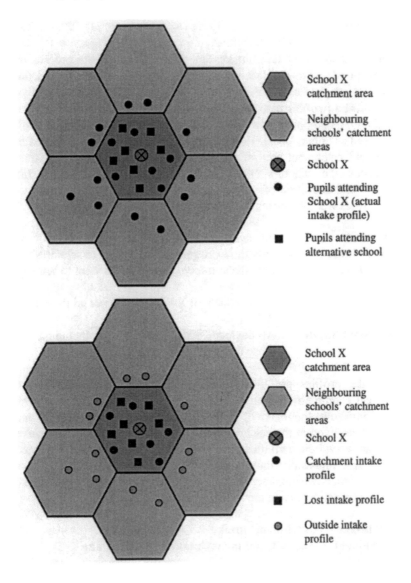

Figure 5.5 Hypothetical example of school X and its intake, with the resulting intake profiles

The rationale for these three comparisons is given below.

'Catchment' intake vs. 'lost' intake These two intake profiles made up the 'all catchment' intake. Therefore, if they were significantly different from each other then it was possible to say that the potential 'catchment' intake was being divided into two particular sets of pupils. One stayed at the local school while the other particular set of pupils left, or rejected, that local school. It was not possible to say if the latter group of pupils rejected the local school through their own choice or if they were being denied a place at that local school. However, most admission policies for LEA-maintained schools stated that local pupils had priority in admissions and, hence, at least in the case of LEA-maintained schools, it was possible to say that the 'lost' intake rejected their local school through choice on their part.

'Outside' intake vs. 'catchment' intake These two intake profiles made up the 'actual' intake. Any differences or similarities in the characteristics of these two subsets said something about social assimilation via the choice process of schools. If these two subsets were similar then the 'outside' impact of the intakes would not have changed the overall intake balance. However, there might have been instances where these two subsets appeared to be significantly different. In this case identifying why would also have been important part in understanding the choice process.

'Outside' intake vs. 'lost' intake These two intake profiles were, in effect, replacements for each other. If they were significantly different from each other then it would appear that the pupils who rejected their 'local' school were being replaced by a very different set of pupils with very different socio-economic characteristics. However, if they were similar then it would appear that there were no differences, at least socially and economically, between the households that rejected a particular school and those households who actively chose that particular non-local school.

To compare and contrast the intake profiles of each school the non-parametric Mann-Whitney U test was used to see whether they significantly differed from one another. Hence, the null hypothesis of these tests was that the intake profiles were not significantly different in the socio-economic composition of the pupils in each. The results of this test are presented in Tables 5.7 to 5.14. In each case only the differences that were statistically significant are given, but they do indicate in which direction they differ.

Comparing the intake profiles of each school using the Mann-Whitney U Test made it possible to look at each school's intake in some detail. In particular, this ascertained if the 'active' choosers of a particular school, and those who rejected a particular school, had distinguishing socio-economic characteristics. In order to identify unique choosers of a school the 'outside' intake profile needed to be compared with the 'catchment' and 'lost' intake profiles. If these were significantly different then it was possible to suggest that the parents actively choosing this particular school had particular socio-economic characteristics that distinguished them from other schools' intakes. If the pupils and parents who rejected a particular school had significantly unique socio-economic characteristics then there would have been significant differences between the 'lost' and 'catchment' intake profiles, and between the 'lost' and 'outside' intake profiles.

The Greater Manchester Metropolitan Schools

Overall, there were many significant differences in the intake profiles of all three Greater Manchester schools. There were few differences between the 'catchment' intake profile and the 'lost' intake profile of Albert Randall, largely because the 'lost' pupils chose to attend parallel 'private' schools (Table 5.7). However the 'outside' intake profile did contrast quite markedly with both the 'catchment' and 'local' intake profiles. The 'outside' intake tended to be from rented accommodation, from neighbourhoods with younger families, and had significantly higher levels of poverty. These pupils, who chose to attend Albert Randall, rather than attending their nearest school, generally came from poorer neighbourhoods in that part of the LEA. In particular, they were from the local catchment area of Pembroke. However, these same pupils from Pembroke were in fact from more socially advantaged households than many of the pupils who continued to attend Pembroke. This can be seen in Table 5.8, which illustrates the significant differences between the 'lost' intake and the 'catchment' intake. So, even though it appeared that pupils from relatively poor neighbourhoods were choosing to attend Albert Randall, these were not from areas as poor as those who continued to attend Pembroke.

The example of St Percival's intake profiles was less clear since this was the type of school that did not have a local catchment, even before the reforms to education provision. However, it was still evident that the pupils attending St Percival's from 'outside' the thiessen catchment were significantly more 'advantaged' than pupils who lived nearby. This can be seen in Table 5.9, which shows the higher levels of wealth Indicators for the 'outside' intake against both the 'catchment' and the 'lost' intake. Since pupils who lived in the thiessen catchment of this school probably never considered it as their local catchment school it was perhaps not surprising that there were no significant differences between the socio-economic characteristics of either the 'catchment' or the 'lost' intake profiles.

Table 5.7   Albert Randall's significant differences in the socio-economic composition of different intake profiles

GB Profiler '91 variables	*Significant differences between intake profiles* *		
	'Catchment' vs 'Lost'	'Outside' vs 'Catchment'	'Outside' vs 'Lost'
Younger People	-	15.72	13.67
Older People	-	-	-8.17
Ethnicity	5.58	-	6.53
Privately Owned Housing	-	-42.75	-31.44
Rented Accommodation	-	40.81	29.32
White-collar Workers	-	-58.38	-41.79
Blue-collar Workers	-14.77	-	-8.12
Young Families and Couples	-	46.26	38.16
Older Families and Couples	-	-16.88	-17.21
Wealth Indicators	-	-46.32	-44.01
Poverty Indicators	-	44.46	40.55

* $p < 0.05$ using Mann-Whitney U test. Positive values mean that the first of the paired intake profiles was higher than the second. Negative values mean that the second of the paired intake profiles was higher than the first.

Table 5.8 **Pembroke's significant differences in the socio-economic composition of different intake profiles**

GB Profiler '91 variables	Significant differences between intake profiles*		
	'Catchment' vs 'Lost'	'Outside' vs 'Catchment'	'Outside' vs 'Lost'
Younger People	-	-	13.65
Older People	-	-10.76	-
Ethnicity	1.27	-	-
Privately Owned Housing	-36.85	-	-28.88
Rented Accommodation	38.79	-	32.58
White-collar Workers	-25.31	-	-23.70
Blue-collar Workers	-16.20	8.73	-
Young Families and Couples	14.23	23.52	37.76
Older Families and Couples	-12.41	-	-12.08
Wealth Indicators	-25.22	-	-21.04
Poverty Indicators	43.23	-27.18	16.05

* $p < 0.05$ using Mann-Whitney U test. Positive values mean that the first of the paired intake profiles was higher than the second. Negative values mean that the second of the paired intake profiles was higher than the first.

Table 5.9 St Percival's significant differences in the socio-economic composition of different intake profiles

GB Profiler '91 variables	*Significant differences between intake profiles**		
	'Catchment' vs 'Lost'	'Outside' vs 'Catchment'	'Outside' vs 'Lost'
Younger People	-	19.23	13.42
Older People	-	-58.07	-44.82
Ethnicity	-	-	-
Privately Owned Housing	-	-	-
Rented Accommodation	-	-29.54	-24.55
White-collar Workers	-	56.03	45.09
Blue-collar Workers	-	18.24	11.82
Young Families & Couples	-	-	-
Older Families & Couples	-	-	9.15
Wealth Indicators	-	57.55	45.49
Poverty Indicators	-	-34.51	-23.11

* $p < 0.05$ using Mann-Whitney U test. Positive values mean that the first of the paired intake profiles was higher than the second. Negative values mean that the second of the paired intake profiles was higher than the first.

The West Midlands County Schools

The intake profiles of the two West Midlands schools, Hengrove and Polden, showed little evidence of divergence. This was as expected, since both schools were designated as being in little or no competition with other 'state' schools on any order of magnitude that would have seriously altered the socio-economic composition of their intakes. The only change to Hengrove's 'actual' intake was the lower level of wealth indicators of the 'outside' subset of the intake in comparison with the 'catchment' pupils (Table 5.10). There was even less significant alteration of the socio-economic composition of Polden's intake with open enrolment (Table 5.11).

The Eastern County Schools

A similar picture was seen in the intake composition of the Eastern County schools, Northleigh and Thorndale, where no significant differences were found at all. The third Eastern school, Hartland, did appear to have significantly changed its socio-economic character through open enrolment (Table 5.12). It was rather surprising that there were few changes to the intake composition of Northleigh and Thorndale, since both schools were very attractive to pupils from 'outside' the catchment areas (Table 5.5). Both schools also 'lost' approximately 20% of the pupils living within their thiessen catchment area. However, even though there were significant changes to the number of pupils choosing or rejecting these schools, there were no significant differences in their socio-economic backgrounds. It could be argued that there was little variation in the GB Profiler '91 index scores for pupils across large areas across this LEA. Alternatively, these schools illustrated greater homogenisation of choice across social cleavages in this Local Authority. However, the example of Hartland did indicate variations in the socio-economic backgrounds of pupils choosing or rejecting this school (Table 5.12).

The greatest difference in the intake profiles of Hartland was between the 'catchment' and the 'lost' intakes. However, these changes were rather complex for they suggested that the 'catchment' intake had greater wealth indicators **and** greater poverty indicators. Similarly, the GB Profiler '91 variables for accommodation also showed that the 'catchment' intake were typically from more privately owned housing neighbourhoods **and** areas with higher levels of rented accommodation. This suggested that the pupils from one of these two intake profiles were concentrated from similar socio-economic backgrounds, and that the other intake profile was composed of pupils from more widespread socio-economic backgrounds, both, more 'advantaged', and more 'disadvantaged' than the other intake profile.

Table 5.10 Hengrove's significant differences in the socio-economic composition of different intake profiles

GB Profiler '91 variables	Significant differences between intake profiles*		
	'Catchment' vs 'Lost'	'Outside' vs 'Catchment'	'Outside' vs 'Lost'
Younger People	-	-	-
Older People	-	-	-
Ethnicity	-	-	-
Privately Owned Housing	-	-	-
Rented Accommodation	-	-	-
White-collar Workers	-	-	-
Blue-collar Workers	-	-	-
Young Families & Couples	-14.42	15.16	-
Older Families & Couples	-	-	-
Wealth Indicators	-	-27.13	-
Poverty Indicators	-	-	-

* $p < 0.05$ using Mann-Whitney U test. Positive values mean that the first of the paired intake profiles was higher than the second. Negative values mean that the second of the paired intake profiles was higher than the first.

Table 5.11 Polden's significant differences in the socio-economic composition of different intake profiles

GB Profiler '91 variables	Significant differences between intake profiles*		
	'Catchment' vs 'Lost'	'Outside' vs 'Catchment'	'Outside' vs 'Lost'
Younger People	-	-	-
Older People	-	-	-
Ethnicity	-	-	-
Privately Owned Housing	-	-	-
Rented Accommodation	-	-	-
White-collar Workers	-	-	-
Blue-collar Workers	9.62	-	-
Young Families & Couples	-	-	-
Older Families & Couples	-	-	-
Wealth Indicators	-	-	-
Poverty Indicators	-	-	-

* $p < 0.05$ using Mann-Whitney U test. Positive values mean that the first of the paired intake profiles was higher than the second. Negative values mean that the second of the paired intake profiles was higher than the first.

Table 5.12 Hartland's significant differences in the socio-economic composition of different intake profiles

GB Profiler '91 variables	Significant differences between intake profiles*		
	'Catchment' vs 'Lost'	'Outside' vs 'Catchment'	'Outside' vs 'Lost'
Younger People	-	7.49	-
Older People	-	-18.01	-11.57
Ethnicity	-5.25	-	-
Privately Owned Housing	6.20	-	-
Rented Accommodation	17.44	47-.11	64.55
White-collar Workers	-20.89	-	-
Blue-collar Workers	162.40	-	224.12
Young Families & Couples	-11.63	-	-
Older Families & Couples	4.52	-	-
Wealth Indicators	29.39	-	29.73
Poverty Indicators	6.35	-	-

* $p < 0.05$ using Mann-Whitney U test. Positive values mean that the first of the paired intake profiles was higher than the second. Negative values mean that the second of the paired intake profiles was higher than the first.

Conclusion

This Chapter has begun to outline the characteristics of eight schools and their intakes. The first half showed how the schools from the three different LEAs represent different examples of the type of competition in the market place. They also represented different geographical contexts in which the decision-making process for choosing a school takes place. The second part of this Chapter identified the major differences in the socio-economic characteristics of the intakes and the significant changes to their composition due to the open enrolment legislation. This highlighted the diversity of backgrounds that pupils attending these eight schools came from, and illustrated the diversity of patterns in the socio-economic transformation of intakes due to school choice that provided a useful contrast for the household-level decision-making process in the next Chapter.

6 The Geography of 'Parental Choice'

Introduction

This Chapter examines the process of choice within the education market place. Choice is conceived within the context of the household and contextualised within the geography of the local area and the market place. The analysis of parental choice is based on a survey of 215 parents whose children attended eight schools from across the study area. The characteristics of these schools and their whole intakes were discussed in the previous Chapter.

This Chapter focuses on the decision-making process of the household in choosing a secondary school. It starts by examining the stages to the process by examining the initial set of schools that parents chose from and then considers the different roles of the parents and the children in the process. This examination of stages to the process finally identifies different types of consumers participating in the market place. The Chapter then proceeds to examine the acquisition of information both for the whole market place and for the final choice of school. This is followed by an analysis of how parents went about making their final choice and the factors that helped them determine this choice. The discussion then follows on from these factors by studying in detail the geographical influences on the choice of school. The Chapter then concludes with a discussion on the parents' own evaluation of the process of choosing a school.

The Decision-Making Process

Choosing a school is a very complex process and attempting to discover how that process operates is a very difficult task. There are many parallels between the decision-making process of choosing a school and residential mobility and the approaches to understanding the behaviour of individuals and households in that process (Walmsley and Lewis, 1993; Golledge and Stimson, 1997). Sheth (1974) identified three different approaches to family decision-making:

i.   Macro vs. Micro Approach.
ii.   Descriptive vs. Determinative Approach.
iii.   Attitudinal vs. Behavioural Orientation.

The first distinguishes between macro and micro decisions, in other words, the distinction between the household and the members of the household as primary units of analysis. In this study examination of the decision-making process is undertaken from both approaches. On the one hand, the form of data collection was based around the household as the unit of enquiry, but it also aimed to identify the significance of the different members of the household in the process of choosing a school. (For a similar approach to residential search behaviour in a rural context see Seavers, 1999.) The second form of approach considered by Sheth (1974) contrasted research that observes and identifies the decision-making approach against research that uses determinant criteria in the analysis. The latter form of enquiry has been criticised within school choice research (Bowe *et al.*, 1994), and decision-making generally (Van Der Smagt and Lucardie, 1991). However, this study used both approaches in order to 'triangulate' the responses of parents, so as to ensure accurate interpretation of the process. The third way of distinguishing between research into family decision-making, according to Sheth (1974), was by focussing on attitudes of families rather than their behaviour. This research, while identifying parents attitudes to school choice and education in general, also tried to get parents to chronicle their behaviour throughout the decision-making process in order to identify potential 'stages' to the process.

Another important aspect to this study's approach to the decision-making process was the context in which decisions were made. As Desbarets (1983, p.21) said, when discussing migration, 'explanations of migration that attribute a large place to the motivations and the preferences of migrants have been limited in their usefulness by their implicit assumptions that choice and behaviour are predominantly under the migrants' personal control'. Desbarets (1983) then proceeded to argue that migration needed to be understood against the institutional constraints and external supply constraints. Judson (1990), who identified several major flaws in migration decision-making research, also requested a similar choice-constraint approach. Among the flaws included: the tendency for preferences to be studied in a vacuum; that value was being examined without reference to other, competing, values or constraints; and that utility maximising models do not accurately represent the decision-making process.

In order to overcome some of these deficiencies this study of the decision-making process focused on six elements of choice, which closely represents the chronological order of the decision-making process. The six elements were:

i.   The stages to the decision-making process
       This attempted to identify if there were some general steps, or stages, in which parents went about choosing a school.

ii.   The extent of choice
       This has been traditionally seen as the first stage of any decision, and, therefore, this study attempted to uncover the range of schools that parents eventually chose from.

iii.   The 'active' role in the decision-making process
       This focused on the 'active' role taken when choosing a school, which was attempted by first, comparing households that could have been called 'active' in the process against those that seemed ambivalent to, or 'inactive' in, the process and, secondly, by unravelling the roles *within* the household, particularly at the parent-child interface.

iv.   Information sources
       Intrinsic to the decision-making process is the acquisition of knowledge. The final decision was often due to the type, quantity and quality of information a parent had access to, and the source of that information.

v.   Choice of school
       The fifth element considered here focused on the actual decision made, and asked why parents chose a particular school and what factors guided them to that decision.

vi.   Satisfaction with choice and process
       The final stage to this analysis was whether parents and their children were happy with the decision that they had made, and if they were happy with the way they went about making it. This gave the parents the opportunity to say how they would have improved or changed the process of choosing a school.

These six elements to the decision-making process are now discussed in turn. First, the discussion starts by identifying if stages to the decision-making process were observable, and if the behaviour of parents, themselves, followed such patterns.

Stages to the Decision-Making Process

When the parents were asked to describe how they went about choosing a school very few actually outlined stages to their decision. This was not surprising since the majority of parents believed that they already knew which school they wanted to send their child to before they believed any kind of process began. Only 22% of respondents said that they did not already have a preconceived idea of which school to send their child to. Of course this already highlights the complexity of choice, since, at some point, the parents did choose a particular school, but for some reason they believed that this was not part of the decision-making process. This suggests that there was some kind of *inevitability* to their choices, but this is considered later.

The high frequency of a pre-determined choice of school also went against the relatively high proportion of parents who did not live locally to the school they eventually sent their child to (59.81%). It can only be assumed that there were parents who believed that they always wanted their child to attend a school outside the local catchment area. So, even though they appeared to have been quite 'active' in choosing a school, their decision-making process was actually quite limited and constrained.

Gorard (1997) suggested from research in Wales on the established education market, i.e. the fee-paying sector, that there were three possible steps, or stages, that parents went through in choosing a school:

Step 1   Parents, alone, decide on the type of school.
Step 2   Parents, alone, choose a subset of schools of this type from which to choose from.
Step 3   Parents and child come to an agreement about one school from the subset.

Gorard then calculated that these three stages could be used to produce five different combinations, or five types, of consumers. Using the simple matrix in Table 6.1 the different levels of engagement could be identified.

Table 6.1 Consumer types as identified by the stages that parents participated in when choosing a school

Step 1	Step 2	Step 3	Consumer type
✗	✓	✓	'Eclectic' – chose from all schools
✓	✗	✗	'Fatalist' – LEA makes choice
✓	✗	✓	'Child-centred'
✓	✓	✗	'Parent-centred'
✓	✓	✓	'Consumerist'

Source: From Gorard (1996, p.246)

Using Gorard's proposed model of school choice this analysis of 'parental choice' highlights the differences that may be currently occurring between the way parents choose a school in the 'new' education market as opposed to the parents choosing schools in the more established and more 'private' education market.

The discussion will now try to address these stages of the decision-making process in the context of this research under three key factors that determine these stages. First, it will look at the choice of a 'type' of schooling that parents make before, secondly, outlining the different schools that parents chose from. Another factor that Gorard identifies as determining stages to choosing a school, the relationship between the parents and the child, is then addressed before finally discussing if particular consumer types (Table 6.1) are identifiable in this study.

Type of School

As mentioned earlier, few parents outlined stages to their decision-making process. However, it was possible to infer from the responses if these three stages were at work. Indeed there was plenty of evidence to suggest that choosing the type of schooling was important. As will be seen later the type of schools that parents chose from very much reflected the type of school of their final decision. Comparing the 'private' St Percival's School (STPE) and the 'state' Albert Randall School (ABRA) in the Greater Manchester metropolitan borough best illustrated this. Both schools were located close to each other but there was a clear difference in the *type* of schools that the parents chose from. The parents with children attending St Percival's School generally chose from 'private', and in particular, Roman Catholic schools, whereas the parents with children at Albert Randall School generally chose from LEA-maintained schools. This was supported by some of the statements made by parents:

> The school is the nearest non-Catholic high school (ABR23).

> Wanted him to be state educated (NOR19).

Interestingly, none of the parents with children attending St Percival's School explicitly said that they chose from 'private' or Roman Catholic schools. However, this was implied when they commented that St Percival's was the closest school, yet, based on their postcode location, there often was a closer 'state' school. Also there was a great tendency for these parents to say that attending St Percival's School was inevitable since their child had attended a feeder Roman Catholic primary school. In such cases the *type* of schooling they wanted for their child had already been decided several years previously.

In the examples where parents did mention which schools they chose from there was a reluctance to choose from across **all** types of schooling – 'eclectics' according to Gorard (1996). In fact there was only one case where this was mentioned:

> St Percival's is the feeder school from St Marie's and was the natural choice. It has an excellent reputation. We did consider private education but my daughter wanted to go to St Percival's with all her friends (STP6).

The reluctance given to this is expressed by the following respondent, which verged on the fatalistic nature of the choice process:

> There was an obvious choice of state school. We are in the catchment, so easy... Independent school considered – but not a serious contender except in unlikely event of getting a scholarship (NOR2).

The Considered Schools

The second stage that Gorard (1996) observed from the decision-making process was for parents to choose a subset of schools. From this study there was a good deal of evidence to suggest that subsets of schools were used.

Parents were asked to list all the schools that they considered throughout the decision-making process. The resulting number of schools that they each indicated varied from just one school (i.e. their final choice) to seven schools, in a small number of cases. Figure 6.1 shows the distribution of the number of schools considered. The largest proportion of all respondents (40%) never considered any school other than their actual choice. This meant that approximately 60% of the respondents considered at least two schools or more during the decision-making process.

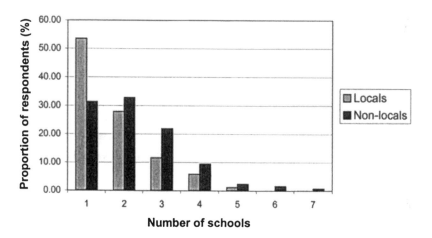

Figure 6.1   Number of schools considered during decision-making process by 'locals' and 'non-locals'*

*   As defined by pupils attending their nearest school or an alternative to their nearest school.

The distinction between parents who did not consider any other school and those that did was very important. There have been many attempts in the literature to identify this. For example, Adler *et al.* (1989) distinguished between 'choosers' and 'non-choosers' while Ball *et al.* (1995) identified *circuits of schooling* that differentiated between 'locals' and 'cosmopolitans'. More recently, Carrol and Walford (1997) suggested that there was a continuum between being 'active' in the market place to being 'passive' in the market place. Therefore, it was necessary for such distinctions to be used in this study. For the purpose of this discussion a simple distinction between those parents who didn't consider more than one school (the 'inactive') and those who considered at least two schools (the 'active') provided a useful framework in which to proceed.

Of these 'inactive' parents only 53% lived locally to their school. Consequently, 47% who only considered one school attended a 'non-local' school. This meant that a good number of parents never considered their local catchment school. It should be noted, however, that the majority of these 'non-local' parents attended the Roman Catholic school of St Percival's, and, as discussed in the previous Chapter, 'private' schools tended not to have catchment areas. Therefore, parents who wished to send their child to a 'private' school did not go through the decision-making process within the context of living in a catchment area or having a 'local' school. Of the St Percival's School parents who indicated that they did not consider more than one school, 96% actually lived closer to another school.

Overall, the low number of schools considered within the decision-making process matched the small number of schools within the *competition spaces* as identified in Chapter 4. Most of the hierarchies discovered never exceeded beyond two or three schools in direct competition with one another.

With the exception of parents from St Percival's School, the parents attending a 'non-local' school tended to consider more schools than those parents who lived within the 'local' catchment area of their chosen school (Figure 6.1).

It was assumed that a 'non-local' parent would have considered more than one school since the 'local' school would have featured quite heavily in the decision-making process. This also supported the case for using a 'local'/'non-local' distinction between parents. It also gave an idea of the accuracy of using thiessen polygons to define the 'local' catchment areas.
This trend between 'locals' and 'non-locals' also appeared to transcend the urban/rural divide. By examining the schools individually a pattern did emerge showing that parents from schools that were more isolated in the market place did not consider as many alternative schools as those parents from urban schools.

The discussion now examines the actual schools considered and attempts to explain these patterns. Wood (1992) defined the 'extent of choice' as being a function of 'objective' factors, such as distance from home and oversubscribed schools, along with parental views and perceptions, and resources. In this study there was a propensity for those parents who considered subsets of schools to define their subset heavily by locality. For example:

Viewed all schools in the area, chose the best one (ABR31).

Choosing was related by the area in which we lived (HAR22).

Discussions with other parents to establish reputations of local high schools (HAR4);

We visited all secondary schools in the area with our daughter... (HEN9).

Looked at the available schools within a reasonable radius of home (POL22).

The emphasis on the local area was certainly reflected in the schools that parents considered throughout the process. One aspect that arose from assessing if the stated schools corroborated the competition spaces was that schools that did not directly compete, but were connected by linked hierarchies, were regularly considered. This suggested that, therefore, the competition spaces should really be seen in their entirety, including all the connected hierarchies.

The schools listed by the parents from St Percival's School reaffirmed what was stated previously in Gorard's (1996) first stage to the decision-making process. All the schools mentioned here were all 'private' schools, four of which were Roman Catholic schools. This compared favourably with the schools considered by Albert Randall parents, who were spatially from the same pool of consumers. In this example, a much greater number of parents were choosing from 'state' schools. However, this example began to highlight the impact that parallel markets had on one another. Parents from both Albert Randall School and Pembroke School listed 'private' as well as 'state' schools. This 'overlap' of schools appeared frequently down the list of all schools, the only exception being parents from Thorndale School in the Eastern county LEA. Consequently, it was difficult to assume that all parents viewed the market place as a series of parallel markets, even if competition between schools reflected this.

There was some evidence to suggest that parents were choosing a school for their child with some idea of a hierarchy in their minds. There were very few occasions when parents considered schools at the bottom of a local hierarchy, and none considered schools more than one level away from their final choice. It could be inferred from this that parents were positioning themselves in this hierarchy.

Only 15% of the respondents indicated that they made a shortlist of schools before finding out more about them. This did rise to nearly 26% of those respondents who did consider more than one school, the 'active' choosers. What seemed to be of more importance to the way parents went about choosing a school was the significance of the 'local' school on the process. Adler *et al.* (1989) found that the avoidance of the 'local' catchment school was important for the majority of 'active' choosers in their study. According to the responses in this study over 70% of the 'active' parents highlighted the importance of the 'local' school in the process. This broke down into 10% who said that they rejected the 'local' school before considering other schools, while 60% said they used the 'local' school as a benchmark when considering other schools. This suggested that the quality and reputation of the 'local' school drove the decision-making process for the majority of parents. Consequently it was difficult to suggest that parents went about selecting schools for consideration in, what Gorard (1996) would call, a consumerist fashion. The small number of schools considered by parents in the decision-making process also supported Adler *et al.* (1989) when they argued that parents were likely to find a satisfactory alternative to their 'local' school, rather than choosing from a wide-ranging choice of schools. However, this did not deflect away from a significant number of parents who did behave in a stronger consumerist way and considered each school equally.

By identifying the schools considered during the decision-making process the discussion has focussed on 'active' parents and 'inactive' parents in the process. This leads to the third characteristic of Gorard's (1996) three stages; who **within** the household was most 'active' in the decision-making process.

Active Role within the Household

Little research has considered the many perspectives of the 'family' in their decision-making processes (Seavers, 1999). However, a significant contribution to our understanding of family decision-making has been the work of Sheth (1974). For example, he proposed three ways in which the family could be 'consumers': individual members of the household as 'consumers', the family acting and behaving as a single 'consumer', and the household unit as an *indirect* 'consumer'. Sheth (1974), subsequently, proposed a theory of family buying decisions, which was based on four key components:

i.　　Individual members of family, their predisposition, and underlying cognitive world of buying motives and evaluative beliefs about products and brands.

ii.   Determinants of the cognitive world of individual members, which are both external and internal.
iii. Determinants of autonomous versus joint family decision-making.
iv.   The process of joint decision-making, with consequent intermember conflict and its resolution.

Of these four, the most important in choosing a secondary school is the inclusion and exclusion of different members of the household in the whole process or at particular times during the decision-making process. Sheth (1974) identified a number of facets that affected the distinction between autonomous and joint family decision-making, including class and role orientation. Research into decision-making has tended to focus on the distinction between just a single parent in the household and dual parent decision-making. So, for example, Haas (1980) identified six ways of identifying the traditional segregated roles of couples: breadwinner; domestic; handyman; kinship; child care; and major/minor decision-maker.

These different roles of the adult members of the household highlight the power relationship between these individuals. However, the decision-making process for choosing a secondary school can also include the child, and, consequently, any analysis of the family decision-making process, must include all members of the household.

A very important element to Gorard's (1996) analysis of stages to the decision-making process was the role of the child. In relation to the three stages of the process it was generally only at the final stage that the child was introduced as a key actor. However, Gorard did not assume that the child, therefore, played a secondary role, since it was suggested that some households would jump from the first stage to the third stage, giving the child a significant input into the decision. These 'child-centred' decision-makers contrasted quite starkly from those households where the child played almost no role at all. Other research has calculated the proportion of households in which the children played varying roles of importance in the decision. Woods (1992) suggested that, from the PASCI (Parental and School Choice Interaction) survey, 80% of the respondents said the responsibility between parents and children was shared, and that only 2% allowed the child to make the whole decision. Rooke (1993), who asked children who they thought had responsibility for the final choice of school, uncovered fairly similar findings. In this case 74% of the children said that it was a shared decision, 10% said that they, the children, had total responsibility for the decision, and 16% said their parents were the only ones involved.

In this study parents were asked as to what proportion of the decision making involved the key household actors - usually between the mother, father and child. Only one respondent indicated that the child had total responsibility for the decision, while 29 parents (14%) said that the child had no role at all. This was in keeping with the findings of Woods (1992), who similarly asked parents, rather than the children, for their perception of responsibility. However, it was significant that, for the remaining 86% of respondents, there were varying degrees of responsibility between the three key actors in the household. Reay and Ball (1998) have suggested that the role of the child in the decision-making process varied according to the social-class of the household. It was argued that children have greater power to influence choice in working-class families than in middle-class families where the parents produced what the authors called a 'veneer of democratic decision-making... [which] ... masks tight parental control' (Reay and Ball, 1998, p.443).

There were even greater variations in the roles of the two parents in the decision-making process. Research by David (1997) and Reay and Ball (1998) has highlighted that the mother usually took the main responsibility for choosing a school, which was borne out by their leading role in doing the groundwork for informed choice. However, Reay and Ball (1998, p.443) did not assume that this meant the mother took charge of the decision, but instead mothers took charge because they are the 'labourers of school choice'. They also suggested that this crossed social-class boundaries.

Results from this survey tended to support both sets of conclusions given above. The average proportions of divided responsibility for school choice given by all the respondents can be seen in Table 6.2. This highlights the greater importance that the mother took and the almost equal role of the father and the child. It also highlights the fact that the parents, collectively, had a much greater say in the choice of school than the child.

For greater analysis it was simpler to look at all the respondents according to general patterns of decision-making within the household (Table 6.3). Clearly the most dominant feature was that in four out of every five households the parents dominated the choice of school. However, 12.9% of the responses said that the three key actors had equal responsibility, and 20.6% said that the child had a greater share than each of the parents, but that collectively the parents still dominated. Obviously, in the latter case it was not assumed that the parents always acted together and chose the same schools, but the detailed literature on this parent-child interface would suggest that their decisions are very closely entwined. At the other end of the scale, only 6.2% of the respondents said that the child had a majority share in the final decision.

Table 6.2   Average responsibility of members of household in choosing a school

Member of household	Average proportion of total decision (%)
Mother	40.86
Father	30.40
Child	28.01

Table 6.3   Variations in responsibility for determining school choice

Share of responsibility for decision	Child's share (%)	Proportion of respondents (%)
Parents had majority share	<50	80.9
All equally responsible	33	12.9
Child had greater responsibility than individual parents but not the majority share	33-50	20.6
Child was responsible equally to both parents	50	12.9
Child had majority share	>50	6.2
Both parents had equal share		64.9
Mother had greater responsibility than Father		28.9
Father had greater responsibility than Mother		6.2

Dividing the respondents into these key roles taken when choosing a school, and looking at their respective socio-economic profiles, highlighted a number of things. For example, this showed that those parents where the child had been given the majority share of the decision were more likely to be from low-income households. As the relative wealth of the households increased then the role of the parents began to equal that of the child, and then collectively began to dominate the decision. But, households in which the parents totally dominated the decision were more likely to have **lower** relative income than this previous group of parents, whilst remaining above the overall average income levels.

Another significant difference between these two sets of respondents was in their education. Households where the child was more important than the parents together were more likely to have been educated in comprehensive schools, whereas the 'parent-dominated' households were generally higher educated and were more likely to have been educated privately. However, parents with the **greatest** and **highest** level of education said they took a more equal share of the decision between themselves and their children.

These emerging patterns of decision-making can be shown diagrammatically (Figure 6.2). This illustrates that as the social 'advantage' of households increased there was a general decrease in the role and participation of the child. However, it also shows that of the most 'advantaged' households there were some variations in the importance they gave their children to the decision. These variations were seen by the distinctions between levels of education and levels of income; the higher educated tended to give their children a greater role, whereas the wealthiest tended to put little emphasis on their children's perspective.

This distinction adds greater complexity to the findings of Reay and Ball (1998), yet supports their key point that working-class households tended to allow their child greater influence over the decision. This may have been due to more limited choice available to these families as much as their inclination to empower their child in the decision-making process. As indicated, the main difference in this study was that the most *educationally* 'advantaged' households appeared to have a more 'democratic' system of sharing responsibility between the key actors. However, this could have represented the 'veneer' that Reay and Ball found in middle-class decision-making.

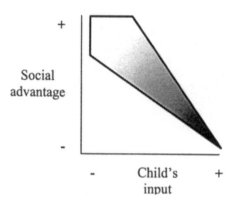

Figure 6.2   Relationship between social 'advantage' and the child's input into the decision-making process

The other key observation from Table 6.3, of key variations in the responsibility of household members in the decision-making process, was the distribution of influence **between** the parents. For the majority of cases, 65%, the two parents shared the responsibility, but 29% of the respondents highlighted the greater importance of the mother in the decision-making process. Households in which the mother was the dominant actor in the process were more likely to be characterised by having low or no qualifications and living in rented accommodation. This contrasted quite dramatically with households in which fathers were the dominant actors, for they tended to have higher qualifications, were university educated and had relatively higher incomes (Figure 6.3). This would suggest that, contrary to David (1997), the role of the mother was linked to the social class of the household.

Consumer Types

At the beginning of this section the work of Gorard (1996) was used to suggest that there were three stages to the decision making process for choosing a school, and that they were largely related to the roles of the key actors within the household. The culmination of Gorard's work was to highlight different levels of engagement (Table 6.1). Even though it was difficult in this study to find evidence of parents perceiving *stages* to the process there was plenty of evidence to support the different types of 'consumers' that Gorard described.

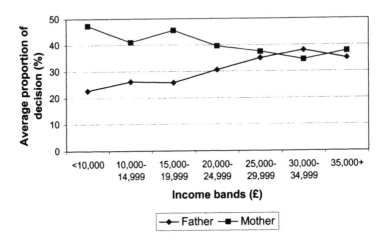

Figure 6.3 Importance of mother and father in decision-making process, by income band

As discussed above, the contrast between 'child-centred' and 'parent-centred' consumer types was quite marked. As one would expect many of the parents mentioned the child in the process, but there were cases where the child clearly had a defining role. There was also some variation in the point of the process at which the role of the child became more pivotal. For example, this varied from coming towards the end of the process:

> We considered the reputation of schools in the locality. We applied to send our son to the Rawbridge School in Tampleton. However, before we were informed that there was no place for him at Rawbridge, our son told us that he did not want to go there, but to go to Polden as all his friends were going there (POL10),

to being central throughout,

> Louise wanted go to Albert Randall, so we looked at the school prospectus and asked parents of children already there. We then went to look at the school, we then left it to Louise to decide (ABR4).

These differed from the 'parent-centred' decision-makers, who tended to focus on what they thought would be best for their child:

After considering my child's requirements and potential I attempted to match up the most suitable school from the resources available (STP1).

The significance of choosing a school of a particular type was very strong in this study. As mentioned earlier there were some overlaps in the *types* of schools considered in the market place. However, few parents came across as being particularly 'eclectic'. Even when schools of different types were mentioned, this was overridden by a sense of fatalism:

> St Percival's is the feeder school from St Marie's and was the natural choice. It has an excellent reputation. We did consider private education but my daughter wanted to go to St Percival's with all her friends (STP6).

The 'consumerists' were much more evident but were still heavily constrained by the geography of the schools:

> We made a list of every possible suitable school. We discarded any which had a bad reputation for discipline. We visited each school's open day and collected as much written info as possible. We applied to all knowing we were likely to be accepted only by one or two due to distance and local demand. My child made the final choice between the two schools she was accepted by, with travel being carefully considered (HEN2).

The 'fatalists', as Gorard termed them, were also very evident in this study and interestingly nearly all of them attached constraints to the process. But, very few actually thought that the decision was really out of their hands. Consequently, many saw the choice of school being linked to the feeder system of schools or to the organisation of catchment areas:

> The primary school he attended was St Percival's feeder school so I had no choice in the matter (STP10);

> I didn't have to choose what school my son was to attend. It was automatic procedure from primary school to high school as a result of the catchment area (HAR10).

Other 'fatalists' stated that since they, or their other children, had attended a particular school, then it was inevitable that their next child was also going there:

> It was taken for granted really that Alex would attend Albert Randall because his sister five years his senior had attended and had been very happy and also received a good standard of education (ABR25).

In these cases the 'fatalism' was a product of a previous decision that had been made and that little attempt was made to readdress that decision. This approach to choosing a school was clearly lacking in economic rationality.

Irrespective of who dominated the choice of school within the household there was only one case in which difference in opinion was mentioned:

> The school that my daughter was to go to was Theobold's as it was her brothers school, but my wife wanted her to go to Pembroke as most of her primary school friends were going there (PEM10).

It was worth noting that in this case the choice of school by the mother prevailed.

Overriding these consumer types was the great value parents attached to having a school **recommended** to them. This was often no more than someone else mentioning the school, but the idea of a school being recommended appeared to add an official seal of approval, and therefore, parents had a considerable urge to use this in explaining how they chose a school:

> The school was recommended by numerous people so I followed up these ideas. I found out for myself at this stage that the school was a very good school and excellent for exam results and this was particularly important to me and my child (STP33).

In most cases the recommendation came from other parents, which highlighted the growing empowerment parents had in choosing a school. If a parent had already been through the process, then they were seen as 'knowledgeable', irrespective of how they went about choosing a school. Whether **all** parents who had been through the process were seen as providing such 'official' recommendations to other parents remains to be seen.

Acquiring 'Knowledge'

This section focuses upon the information, and its sources, used in the decision-making process for choosing a school. Bowe *et al.* (1994) has talked about the use of the 'landscape of choice' as a flexible conception for both, the context, and the way parents build a 'picture' of the available school choices they have. Consequently, the first part of this discussion attempts to address the way parents tried to get a 'total picture' of the education market place, or their 'landscape of choice'. It also examines the sources of information used by parents in their approach to build their 'landscape of choice'. The second part of the discussion focuses on the different types of information sources used by parents to make the decision about which school to send their children. Because both parts of this discussion examine the sources of information there is some repetition. However, the first part of the discussion only considers a few key sources of information, and how they were used by parents, alongside the schools considered in getting an overall 'picture' of the market place. The second part of the discussion provides a greater focus on the sources of information themselves.

Researching the Education Market Place

Parents were asked to specify which, of a predefined set of statements on collecting information, they agreed with. This identified whether the collection of information and acquisition of knowledge was, relatively, planned or unplanned. Of all the respondents, 35% agreed that they had planned this, while 31% agreed that there was little or no planning, and 21% agreed that they didn't collect any information at all.

The proportion of respondents who said they did not collect any information varied considerably by school, but this was related to the level of 'inactive' parents at each school (Figure 6.4). There were three schools that over- and under-represented the number of respondents who collected no information. Hartland School had a very high proportion of its parents who did not collect any information, whereas Thorndale and Pembroke, which had a similar proportion of parents who were 'inactive', had a significantly low level of parents that did not collect any information. The difference between these two cases was the number of parents who did not consider more than one school, but still got information on the school they wanted their child to attend. This was perhaps not unexpected in the case of Pembroke parents, since they lived in a well-developed market place, with a good deal of market activity, or choice, occurring across the whole LEA.

The more obvious contradiction arose in Thorndale and Hartland Schools, which, though located in the same Local Authority and both in very rural locations, there were slight differences in their levels of relative activity when it came to acquiring information.

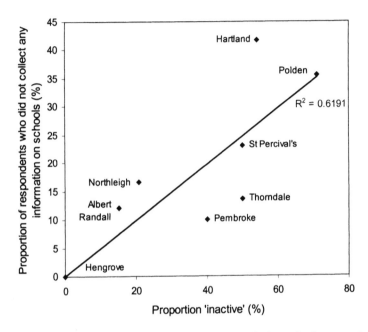

Figure 6.4 **Relationship between the levels of 'inactive' parents* in each school and those who did not collect any information on schools**

* As defined by the number of schools considered (0 = 'inactive').

Turning now to those parents who did collect information, the proportion that planned this collection was almost equal to those that did not. However, there were some interesting differences between schools. Both Pembroke and Polden Schools had a relatively high proportion of parents who agreed that they did not plan to collect information. But these were schools with a high proportion of parents who did not collect any information at all. It seemed, therefore, that for schools where there was a generally low level of market activity by parents then those that did collect information never really planned to do so. This can be seen in Figure 6.5, which shows the almost direct inverse relationship between the proportion of 'inactive' parents and the proportion that planned to collect information.

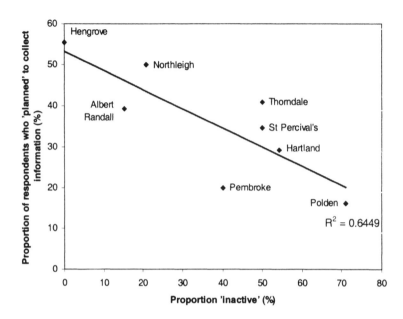

Figure 6.5 Relationship between the levels of 'inactive' parents* in each school and those who 'planned' to collect information on schools

*   As defined by the number of schools considered (0 = 'inactive').

Parents from Thorndale School did not follow this trend, with a relatively high proportion of them having planned to collect information even though they did not really consider more than one school. Clearly these parents had the inclination to acquire knowledge but did not have the capacity to do so. In particular, they were constrained by a lack of choice.

Information Sources Used in the Decision-Making Process

The discussion now considers the information sources used to acquire knowledge on the schools contained in the parents' 'landscape of choice', or schools that they were realistically choosing from. Across all the respondents there were 373 schools considered and parents were asked to indicate which of four pre-defined information sources were used for each school.

Table 6.4 shows how often these four sources were used for the total 373 schools considered. School prospectuses were most frequently used, followed by visits to schools and consideration of their examination results. The least used source of information was from meeting the headteacher of the respective school. These variations clearly reflected depreciating levels of access to these sources. However, there were some interesting observations to be made when considering these sources alongside the number of schools considered by each respondent. Figure 6.6 shows that school prospectuses tended to be used by a similar proportion of parents, whether they were *slightly* 'active' or *very* 'active' in the market place. On the other hand, the more 'active' parents used the three other sources more frequently. In particular, both visiting a school and meeting the Headteacher were very dependent on how 'active' the parent was. The more 'active' the parent was in choosing a school the more likely that these sources were used. This is even more significant when one realises that the more 'active' parents in the market place were accessing these sources much more often simply because they were considering more schools throughout the process.

Table 6.4   Use of four pre-defined information sources in acquiring knowledge of the parents' 'landscape of choice'

Information Source	Proportion of all schools considered
Prospectus	86%
Visited school	77%
Examination results	70%
Headteacher	48%

These findings certainly confirmed that access to particular sources of knowledge was intrinsically linked to the ability (inclination and capacity as Gewirtz *et al.* (1994) calls it) of the parents to interact with the education market place. And these particular sources of information happen to be characteristic of 'hot knowledge' (Ball and Vincent, 1998). It could be argued that a parent was likely to get the most accurate image of a school from 'hot knowledge', such as visiting a school or meeting the headteacher, rather than by reading a prospectus and looking at the examination results. They are, therefore, more likely to accurately match the school to the requirements of their child from these sources. Consequently, having the ability and inclination to research the education market place did mean that different knowledge was acquired.

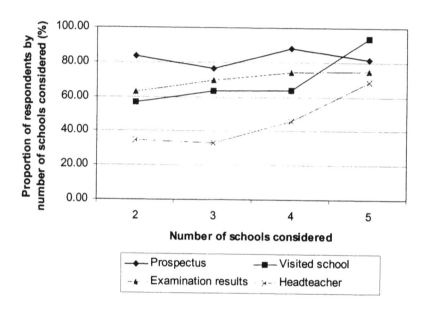

Figure 6.6 **Information sources used by parents, by the number of schools they considered in the decision-making process**

The sources of information most referred to when describing the process provided 'hot' knowledge via the grapevine and were based on experience. The source of information mentioned most was by talking to other parents:

> Word of mouth comments from friends and acquaintances was main factor in drawing up a shortlist (NOR14).

It has been argued that most communication between parents is across particular social networks and that these 'local and personal social networks mediate public/private activities like parental choice and are thus crucial in developing an understanding of the practices and meaning of choice' (Ball and Vincent 1998, pp.378-9). Ball and Vincent (1998) considered the 'grapevine' and showed how this contrasted with other, more official, information.

Examining the responses of parents who mentioned other people as sources of information showed that there were various levels of relationships that these parents had with the grapevines of school knowledge. These could be summarised as follows:

i.   Those that *listened in on* the grapevine.
ii.   Those that *engaged with* the grapevine (indirect interaction).
iii. Those that *used* the grapevine (direct interaction).

At one such level many parents said that they only 'heard' or 'listened to' particular opinions, or views, about schools. This would suggest that there was limited interaction with the grapevine, and was generally, therefore, one-way. So, they said, for example:

> I was quite happy to send him to Thorndale, I had heard a lot of good things about the school from parents of current and previous pupils (THO3).

Another interpretation of the types of relationships that parents had with the grapevines was that they were much more discursive, or two-way. Parents said, for example, that they:

> Discussed experiences of parents with children attending various schools (ABR20).

This, more engaging, relationship with the grapevine can also be seen as either direct or indirect. Indirect interaction was when parents simply said that they 'talked to' other parents. For example, they:

> Looked at the available schools within a reasonable radius of home and talked to parents and children already involved in [those] schools (POL22).

These examples contrast markedly with parents who directly engaged the grapevine and referred to 'asking' the grapevine for information. Therefore, instead of the ambivalent discussion some parents had, there was greater purpose to these discussions. Hence, they got their information:

By asking other parents about schools in the area (THO16).

Ball and Vincent (1998, p.381) mentioned that access to grapevines they studied was 'structured primarily by class-related factors. Where you live, who you know and what community you belong to are vital determinants of the particular grapevine that is open to you'. The level of interaction discussed here also seemed to be related to class. The parents that appeared to *use* grapevines to acquire knowledge directly, were generally higher educated and had larger incomes. As the level of interaction became less direct then the social class of the parents also appeared to fall. So, the parents who *listened in on* grapevines had the lowest incomes and were the least likely to be higher educated. The differing levels of engagement could be equated to differing access to the grapevines, but it could also have indicated how much parents wanted to use, or to be seen by others to use, and engage with, their social networks in the decision-making process.

Finally, one more important feature of the grapevines regarded the importance of the 'locale' on the grapevine. Many parents referred to local opinion, or the views of other parents in the area. Therefore, the acquisition of knowledge through social networks must be seen in its geographical and bounded sense.

It was quite notable that, where the grapevine was spatially defined, the level of engagement with the grapevine was fairly limited. It could, therefore, be argued that social networks for the more working-classes, who were reluctant, for whatever reason, not to engage fully with the grapevines, were more local in character.

The second key to acquiring 'hot' knowledge was based on experiences of the schools concerned. These 'experiences' came from three different sources. The first was generally the source of all grapevine knowledge, that of the experiences of other parents' children. But two other sources of experience also need mentioning since parents frequently mentioned them. The more significant of these two were based on the experiences encountered, and learnt, from having another child at the school:

> My two elder daughters went to Pembroke High and I was very pleased with the attitude from teachers towards pupils and parents (PEM1).

Parents acquired the other source of 'experienced knowledge' mentioned when they were at school:

>I attended Polden High School and considered it the best place for my children to attend. Even then the standards were high (POL24).

It is often taken for granted that 'hot' knowledge is more *immediate* (Ball and Vincent, 1998), but clearly these experiences were at least a generation old and could not necessarily be that accurate. If the market place being researched included the school(s) that the parents attended, which it did in a number of cases, then when studying the acquisition of knowledge it must be noted that the experiences of the parents will be very influential in the final choice. 'Reputation as the basis of judgement and choice, appears to lag behind the actual developments within the schools by many years' (Glover, 1992, p.229). These experiences, even though they may be as powerful as other's experiences, were not, necessarily, going to provide an accurate portrayal of schools many years later.

Choice of School

The final stage in the decision-making process was the actual process of choosing a particular school. This section now considers this final stage by initially looking at the evaluation of the considered schools and then looking at the factors that parents looked for in their final choice of school.

Evaluating the Choice of Schools

Respondents were asked to describe the way in which they evaluated or chose from their considered schools. From their responses it was seen that there was little variation in the way parents evaluated the choice of schools. The smallest number of parents (15%) agreed that they had compared all the schools equally. This distinguished parents who believed that they really did give equal consideration to schools from parents who *attempted* to, and, therefore, probably accepted that there were biases in their evaluation (25% of respondents). At the other extreme, 19% of respondents agreed that they did not give equal thought to each school during the final stages of the process and never attempted to. However, the majority of parents agreed that they had good reasons to reject other schools.

There was also a positive relationship between parents who considered that they attempted to give equal consideration to all schools and their degree of activity in the market place. The parents that agreed that they never gave equal consideration to schools were largely 'inactive' parents, and the proportion of parents dramatically fell with their increasing levels of market activity. Parents that said they had good reasons to reject other schools were unevenly distributed among the different levels of activity in the market place. However, a significant proportion of the 'inactive' parents used this approach, which suggested that even though they never formally considered other schools a good number were still comparing schools. Rejecting schools was the most frequently cited approach for those parents who only considered two schools. This supported the findings of Adler *et al.* (1989) who said that the majority of parents in their study tended to choose a satisfactory alternative to their rejected local school. This would suggest that parents who only considered two schools were only considering a satisfactory alternative to their local school and were, therefore, basing their decision on the rejection of a particular school rather than comparing them equally and side-by-side.

The difference between the parents who agreed that they did compare schools equally and those who *attempted* to give equal consideration was in their levels of education that they had received. Those that said they *attempted* to give equal consideration were better educated and had higher academic qualifications. Those that said they *did* give equal consideration had much lower academic qualifications. Two conclusions were drawn from this. The first was the more educated parents possibly had underlying biases in the final stage of the decision-making process. This could be because they came to this point with ideas or perceptions of what was the 'best' school in the area, irrespective of what they had found out throughout the process of acquiring knowledge. So, it may be that the reputations of schools were much more prevalent in the minds of the more educated parents than it was in the minds of the less educated parents. The second conclusion was linked to the honesty of the parents in the way they evaluated the schools. It might be said that the less educated parents were less likely to be aware of their own preconceived ideas of the schools and, therefore, did not give equal consideration to them.

Factors in Final Choice of Schools

Respondents were given a list of 25 different factors, ranging from being of no importance to being extremely important, to rate, in their final choice of school. Figure 6.7 shows the average responses for all these pre-defined factors. These factors could be categorised as, either, *product criteria* – based on outcomes of the school (e.g. examination results) or *process criteria* – based on human relationships (e.g. the happiness of the child). This distinction was first used by Elliot *et al.* (1981) but has been repeatedly used (for example Stillman and Maychell, 1986; Adler *et al.,* 1989). The majority of these studies have argued that *process criteria* were dominant in explaining parents' choices of schools. For example, Stillman and Maychell (1989) showed that from 7,689 different reasons given for the final choice of school, 68% were *process criteria*, 18% were *product criteria*, and 14% were geographical.

Of the five most important factors that emerged from the replies in this study, four were process criteria. The most important factor overall was the need for 'a good learning environment', followed by a 'caring environment', 'good management' and a 'happy environment'. On the other hand, the most important *product criteria* was a 'strong policy on discipline', followed by 'good facilities in school', 'plenty of resources for students' and 'safety in school'. The next two factors, in 9th and 10th position respectively, were a 'good reputation' and 'examination results'.

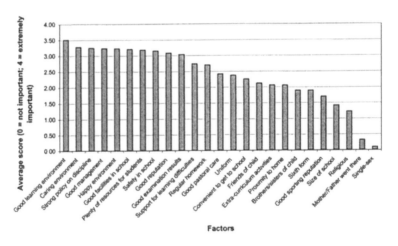

Figure 6.7   Importance of pre-defined factors for choice of school

Before being asked to rate the factors involved the parents had already outlined verbally why they chose a particular school. The two most important reasons given were the reputations of the schools and their examination performances. So when asked why they chose their particular school parents replied:

> Because it had a very good reputation, exam results and friendly environment (HAR10).

These two factors were followed by 'discipline', then 'having a sibling at the school' (19th highest in list of factors), 'proximity' to the school, and whether the schools were 'caring' or not. The only factor that was placed in a similar position using either method was the importance of discipline. Otherwise, the open-ended responses were very different to the ratings of the listed factors, to such an extent that few process criteria were mentioned in the unprompted responses. This raised methodological issues concerning the way these results should be interpreted. However, it could still be said that the rhetoric of 'parental choice' is still dominated by reputations and examination results. It is these two factors that parents consistently used to define their choice of school. Conversely, it could be argued that, on balance, there were other factors that were more important, or, what parents wanted to be more important.

Despite the findings from the pre-defined factors for school choice, the widespread coverage given to the importance that examination league positions and reputations played in the final choice of schools, confirm some of the earlier conclusions. First, the role of examination performance criteria corroborated the findings of Chapter 4. It was suggested then that the more 'state' schools were generally ordered in a hierarchical competition space that was related to the individual school's relative examination performance. Secondly, the reputations of schools still dominated the minds of parents when choosing schools for their children. It is possible to suggest that these reputations were intrinsically linked to the social networks, or grapevines, discussed earlier, and were not necessarily real or accurate reflections of a school.

The other factor that was mentioned quite often in the unprompted questions but not as significant in the list of ratings was the importance of geographical factors. Proximity and accessibility were both used as ways of defining the final choice and could, therefore, not be underestimated. Consequently, many parents said the reason for their choice was:

> because of the locality and distance (POL20),

or,

> because it was the nearest Roman Catholic High School (STO8).

Adler *et al.* (1989) believed that, in their Scottish study, geographical factors were more significant than educational factors in the final outcome of school choice. Putting proximity and accessibility together makes them the third most mentioned factor from the open-ended responses.

The Geographical Influences on Choosing a School

This section addresses some of the more explicit geographical constraints upon the decision-making process in two parts. First, it will examine the spatial constraints upon the travelling arrangements to school that hindered the choice of school, and, secondly, focuses on the 'local' and 'community' role a school plays and how that might influence the choice of school.

Travelling to School

Respondents were asked if, when they were choosing a school for their child, there was a maximum length of time and maximum distance for their child to be travelling each day. Table 6.5 shows that nearly three quarters of the parents said that there were time and distance constraints when they were making the decision, and in retrospect, on their choice of schools.

Table 6.5   Did parents believe there was a maximum length of time and distance for their child to be travelling to school?

Maximum...	Proportion of respondents (%)		
	Length of time?	*Distance?*	*Distance in retrospect?*
Yes	74.3	57.9	71.0
No	25.7	42.1	

The parents who said that there were no time constraints were just as likely to be from urban areas as they were from suburban and rural areas. But, they were more likely to be educated in a comprehensive school, have little or no academic qualifications, were living in social housing and were unlikely to have been to university. However, there was no significant difference in their household income compared to the rest of the respondents. These characteristics were similar for those parents who said that there was no maximum distance, except these were more likely to be from rural areas. It was very interesting to see that the parents with the least advantageous education themselves said they were unlikely to place any time or distance constraints on their choice of school. This will be considered further in the next section on constraints in the decision-making process.

Of the parents who believed there was a maximum time and distance for their children to be travelling, 50% wanted their child to get to school within half an hour and within four miles of their home, three-quarters of them within 45 minutes and eight miles. This set the context to their choice of schools and also supported the relatively small competition spaces identified in Chapter 4. The maximum distances proposed by the parents were related to the environment in which they lived. Urban parents generally did not consider the kind of distances that rural parents conceived. Interestingly, the suburban parents also constrained their choice to a relatively small area from their home.

It could be argued that these varying constraints were related to the available choice to parents, i.e. urban parents believed that they were constrained by a smaller area simply because within that area there were plenty of schools to choose from. Therefore, rural parents, in order to get an adequate choice, had more relaxed constraints on the distances they wished their children to travel. However, in terms of travelling *time* the results were mixed between urban, suburban and rural parents. Consequently, the varied constraints of distance were more likely to be due to the unequal space-time dimensions of urban and rural life. So, instead of the earlier suggestion that the number of schools within an area determined the maximum distances imposed on the choice, it was likely that the time it took to get to a school determined the number of schools being considered.

This line of argument is still subject to the transport arrangements made in order to get a child to a school, and it is to this that the discussion now turns.

Table 6.6 shows the importance of particular travel arrangements for all the respondents. However, the importance of each form of travel varied by school and location. For example, children attending the urban schools of Albert Randall, Northleigh and Pembbroke were more likely to walk to their school. Children attending the other urban school, St Percival's School, rarely walked. Instead they went to school by school bus (cost) or by car on an existing journey. This was because this school attracted students from, for an urban school, a relatively wide area (approximately 4km). Polden, the suburban school, also relied quite heavily on its pupils walking to school. Even though the distances were fairly similar to those travelled by children attending St Percival's this type of area was obviously more conducive to walking to school.

Rural schools tended to be more reliant on school buses. Respondents attending either Hengrove or Thorndale used free and charged school buses, yet parents at Hartland, in the same authority as Thorndale, only used free school buses. The difference between these rural schools was that many of the parents from Hengrove and Thorndale did not live locally to these schools. These rural schools also relied heavily on children getting to school by car, but, in stark contrast with urban schools, these were generally special journeys made to the school and back. The difficulty of getting to a school which was part of an existing journey each day must have been much harder in rural areas and, therefore, only perpetuated the already high costs and constraints of getting a child to a particular school.

Table 6.6 Travelling arrangements of parents

Form of travel	Proportion of respondents (%)
Walking	31.78
School bus (cost)	21.50
School bus (free)	18.22
Car – special journey	11.68
Private bus	7.01
Car – existing journey	6.54
Cycling	2.34

Comparing 'local' and 'non-local' parents illustrated how getting to an alternative school to the local catchment school required particular methods of travelling. Figures 6.8 and 6.9 show the average distances travelled by 'locals' and 'non-locals' by the different forms of travel arrangements to urban and rural schools. In rural areas parents relied heavily on school bus routes, both for those travelling to their local catchment school and those that travelled to alternative schools. This generally broke down to 'locals' having free transport and 'non-locals' having to pay a charge. The non-free school buses did allow the parents to send their children further (10km on average) than they would have with a free school bus (7km on average), but this was heavily constrained by the cost and the particular bus routes that already existed. Consequently, 'non-local' rural parents had to use special car journeys to cover long distances (again, 10km on average).

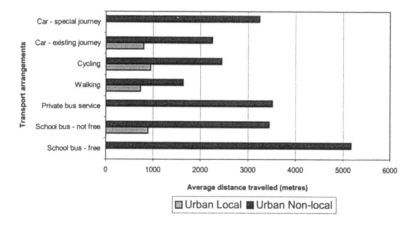

Figure 6.8 **Transport arrangements for urban 'locals' and 'non-locals'***

* As defined by pupils attending their nearest school or an alternative to their nearest school.

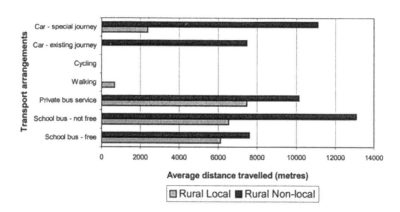

Figure 6.9 **Travel arrangements for rural 'locals' and 'non-locals'***

* As defined by pupils attending their nearest school or an alternative to their nearest school.

Respondents in urban areas were close enough to schools to allow their children to walk to school. Walking was the main way of travelling to school for nearly all 'locals' in urban areas. It was also a key form of travelling for 'non-local' children, up to which, on average, they could go twice as far as 'local' children travelled. However, these distances were still very small (1.6km on average), and probably only allowed parents to choose from one or two nearby schools. So, if parents wished to send their child to a school that was not particularly close then they had to use alternative, and costly, means of transport. Urban 'non-locals' used private bus routes and services to attend schools relatively further away (3.5km on average). Similarly, car journeys, both existing and special, were also used by 'non-local' urban parents to get their children to schools that were, on average, 3km away.

Having identified the forms of transport that enabled parents to send their children to schools other than their 'local' catchment school it was necessary to identify the socio-economic characteristics of those parents who used these different means of transport. There were some very distinct differences in the socio-economic make-up of these parents. The more 'disadvantaged' set of parents got their children to school either by walking or by private bus service. Both of these forms of transport did allow parents to send their child further than they generally would have to get them to their nearest school. However, the extra distances covered by both these means were very limited compared to other forms of transport arrangements.

School buses and car journeys were very much used by the more socially 'advantaged' and affluent section of the respondents. They were generally more educated, had much higher qualifications and had relatively high incomes. It was noted that the more costly of these travelling arrangements, the charged school bus and the special car journey, were used by the least affluent of this set of 'advantaged' parents.

Consequently, there were clear divisions as to which set of parents had access to the travelling arrangements that allowed them to consider schools slightly beyond the immediate locale. This also showed how the least 'advantaged' parents still had access to choice, by getting their child to walk further than they would normally or to take a private bus service to school. However, it was also shown that the potential distances that can be covered by these travelling arrangements were much more limited than they were for more affluent parents.

The 'Local' and 'Community' School

One of the many roles of a school is to act as a key institution within the community. It has been said that the 'greater the degree of community involvement the greater the degree of understanding of the school' (Glover, 1992, p.227). This section examines if there was a greater propensity for schools that were perceived by parents as part of the community to keep their local catchment pupils. This also becomes important if a sense of belonging to a school is dependent on which school the child attends. Consequently, the role of the school to serve a geographically bounded locale as an institution would be eroded away with greater market activity.

This sense of belonging is examined using two slightly different perspectives. One is to consider if a school is perceived to be a *local* school, and the other is to consider if a school is perceived to be part of the *community* that the parents lived in.

Even though nearly 60% of the respondents in this survey were 'active' in the market place, 85% of all respondents still believed they had a local secondary school and 68.7% said they had a secondary school that was part of the community they lived in. On this evidence it was difficult to say that an active market place in education was necessarily reducing the *local* role of a school, but there were fewer parents identifying a school with the stronger, *community*, role.

The schools identified as the *local* school or *community* school were considered using the following Table 6.7. Nearly all of the schools identified as being a *local* school or a *community* school were either the present school or the nearest school. The most notable observation to make from this was that the *community* school was much more likely to be the school the parents sent their children to.

The first question to ask was if attending the nearest school meant that a respondent was more likely to identify with a *local* or a *community* school. Table 6.8 shows the proportion of parents who said they did not identify with such schools.

In this case there were large differences between the expected proportion of respondents and the actual proportion of respondents who did not identify with schools in these ways. Therefore, it could be said that there was a greater propensity for parents not attending their nearest school to state that they did not have a *local* or a *community* school. Once again, this was stronger in the identification of a *community* school.

Table 6.7   Present and/or nearest school as the identified *local* and *community* schools

Identified school as...	Proportion of all respondents (%)	
	Local?	*Community?*
Present school	68.7	82.3
Nearest school	60.3	63.0
Both present and nearest school	42.9	48.9
Either present or nearest school	84.1	95.2

Table 6.8   Parents who did not identify with a local or a community school according to whether they were attending their nearest school

Respondents	Proportion of respondents (%)		
	All respondents	*Had a local school*	*Had a community school*
Attending nearest school	*40.5*	34.4	29.9
Not attending nearest school	*59.5*	65.6	70.1

The three schools with a relatively high proportion of respondents not identifying with a local school were Pembroke (50% of respondents), St Percival's (30%) and Hartland (17%). St Percival's was a 'private' school and, therefore, most of their pupils did not come from the immediate locale of the school since it did not necessarily have a catchment area. Pembroke and Hartland did have more 'local' intakes but both schools were at the lower end of their respective state hierarchical competition spaces, Pembroke being at the bottom and Hartland in the middle of the hierarchies. This would suggest that parents who had children attending these poorer performing schools tended to disassociate themselves from a *local* sense of belonging to them. This observation was further enhanced when examining the nearest schools of the parents who said they didn't have a *local* school irrespective of the school they actually attended.

There was a much greater tendency for parents who lived closer to a school at the bottom of a competition hierarchy to not identify with having a local school. Similarly, a significant proportion of parents with children in schools at the top of a competition hierarchy also did not identify with having a local school. All of these parents had children attending more 'private' schools. Therefore, these parents were likely to disassociate themselves from their nearest school because it was not of the 'type' of school they actually wanted.

The proportion of respondents who did not identify with having a *community* school was much more evenly spread across the different schools, and was not related to the competition spaces of the schools. Even by considering the competition space of the nearest school did not help to define those that had a *community* school from those that did not. However, there was a significant relationship between the average distance that the children travelled to each school and the proportion of respondents who identified with a *community* school (R^2 = -0.65). However, there was no relationship between the distances of the parents from the schools and whether they identified with having a *community* school. Consequently, the conclusion that could have been drawn was for parents who attended schools, that had intakes from a relatively wide area, were less likely to feel a sense of *community* within that school. Therefore, the spatial extent of the intakes was seen to harness, or dissipate, a sense of belonging to the schools.

Evaluating the Decision-Making Process in Choosing a School

The final discussion on the decision-making process of choosing a school examines the parents' thoughts on the choice process in two parts. This section begins by analysing how parents, retrospectively, perceived the way in which they went about choosing a school and how successful they were. It then considers the broader picture and discusses how parents perceived 'others' in the market place and the role that 'choice' had in the education system.

Making 'That' Choice

Nearly all the parents (81%) agreed that they were prepared to make the decision of which school to send their child to. Only 3.23% of the respondents wished that they had made more effort in finding out about other schools, and it was notable that none of these parents attended schools at the 'top' of state competition hierarchies. Even though the vast majority of parents were prepared to make the decision only 60.6% of respondents agreed that they knew exactly what they wanted from a school. 20.2% admitted to having to think long and hard about what would be best.

The impression generated from the responses was that there was a difference between those that were 'inactive', and 'ended up' at a particular school, and those that were 'active' but still chose schools at the 'bottom' of local state competition hierarchies. This latter group, even though they were 'active', usually chose their nearest school. For example, 40% of the respondents who chose Pembroke, a school at the bottom of a local hierarchy, said that they thought long and hard about their final choice of school. This was on a par with the proportion of respondents from Hengrove school who also said this, yet Hengrove appeared to be a far more *attractive* school. In fact, of the 'inactive' parents, those that said they thought long and hard about the choice were all from urban areas. This highlighted the greater sense of market activity in urban areas. Urban parents who were limited in their capacity to extend their choice of schools were still showing signs of inclination that was missing from the equivalent set of parents in rural areas.

There was a general degree of contentment by respondents with the process of choosing a school. This was reflected in the limited number of parents who would have liked more information throughout the process. Of all the respondents, 89.7% said they would not have preferred more information at the time of making the decision. The 10.3% of respondents, who did want more information, were generally the most 'active' in the market place. The most requested type of information that these parents would have liked was:

more information from other schools (ABR19).

Some parents had specific requests such as:

extra-curricular activities; community links (THO14)

and

drugs policy (ABR27).

There were even a few parents who wanted greater honesty in the information that they already had. So, parents would have liked, for example:

> The facility to visit the school unexpectedly to actually see it working – not 'set-up' as at open evenings (ABR12).

There was also a desire for greater 'hot' knowledge rather than 'cold' knowledge:

> Rather than referring to printed information I wish I had spoken more to parents of children at the schools we were considering (NOR24).

More parents thought there was insufficient information on their **rights** in choosing a school (28.5%). Still, the majority of parents (71.5%) were happy with this. Once again, the respondents who wanted more information on their rights were actually the most 'active' in the market place. The difference with the request for more information to help make the decision was that it was clearly the most educated and highest qualified of the 'active' parents who wanted more information on their rights. It was not certain whether these more 'advantaged' parents wanted more information on their rights because they were the most dissatisfied with the process, or whether they were more likely to know they had rights and, therefore, wanted to know more about them.

This led on to whether the parents thought there was a lot of choice between schools in the state education system. The parents were fairly divided on this. The slight majority, 57%, thought there was a good deal of choice, but a significant 43% did not agree that there was much of a choice between schools. A similar figure of 49% of respondents said that they thought there were major obstacles that hindered their choice. These proportions varied by school, and, not surprisingly, it was the schools with 'no competition' competition spaces that had the greatest proportion of respondents who thought there was not enough choice between schools, Hengrove and Polden. However, it should be noted that it was the more 'active' parents at these schools who were the most dissatisfied with the amount of choice between schools.

The schools with the smallest proportion of these dissatisfied parents were Pembroke and Albert Randall, both of which faced the most competition from other schools. Therefore, the size of the competition spaces, identified in Chapter 4, was reflected in the level of choice that parents believed they had.

The greatest reason given by the parents for the lack of choice between schools was the lack of places at popular schools:

> A school only has a limited number of places. When full from the catchment area no one else can get in (NOR8).

This constraint did not necessarily just reduce choice, for some it removed all choice:

> It appears there is no choice really – other secondary schools in our area did not take 'out of catchment' pupils unless you appealed and were still not guaranteed a place. The year my daughter started secondary no 'out of catchment' pupils were taken in to our fairly local school. State schools are too big so they cannot manage to take in 'out of catchment' [pupils] and therefore limit parental choice – in fact often no choice – just state/private (NOR12).

The second most important set of reasons given that limited the choice of schools was geographical:

> I think in our case it is principally geographic based on where we live (i.e. a rural area). It may be that there is more real choice in urban areas with several schools in close proximity (POL10).

It was usually assumed that this was a rural problem and that access in urban areas would not be as constraining. One respondent identified that the reasons for limited choice depended upon whether they were in urban or rural areas:

> In a rural area distance to school is the key criterion. In an urban area the school selects the pupil! (HAR24).

However, this geographic constraint should not be underestimated in urban areas since it was clear that accessibility was a relative phenomenon:

> The only other choice in schools I had within walking distance was Albert Randall and it is a 40 minute walk home. Plus the school entrance is on a busy main road. I have been past the school on many occasions when school has finished to find a great many of the

children are running out into the road creating a dangerous hazard (PEM4).

Procedures were also seen as a constraint on the available choice. In particular, there were a number of criticisms of the catchment area and feeder school system. As one respondent explained:

> It is difficult to get children into schools outside the catchment area you live (HEN4).

The system of education and the allocation of places was specifically criticised in a number of cases:

> Parents should be able to choose a school then fill out a single application form and submit it to that school. Not to be required to write lengthy justifications for not staying within catchment areas (HAR4).

Even then the system was still seen as being problematic:

> The schools with good reputations are over subscribed. If you do not get the first choice, chances are that you do not get in the second choice and end up being 'allocated' a place in a school that you do not wish your child to attend (STP50).

In both of these cases it was suggested that the cultural capital required to overcome these problems, either by writing a 'lengthy justification' or to strategically make your preferences so as to at least get a school of your choice, was considerable.

One respondent also hinted at the hidden social selection of the choice process. So when asked why there was not much choice between schools they cryptically blamed:

> the so called list, and where you are on it (STP36).

But, otherwise, the remaining parents argued that the restriction on choice was linked to the types and diversity of schools. The main criticism was that the schools were too similar. For example:

> In this area the education authority makes sure all schools are of a reasonable standard. There are differences in urban areas but we opted to avoid that problem (HAR7).

Some of these parents saw the development of the National Curriculum as reducing the choice between schools.

On the other hand some parents saw diversity as being equally constraining on the choices available to them. One respondent argued that the main reason for a lack of choice was because of:

> religious schools, i.e. Catholic Schools are not available to children who are non-Catholics and we cannot include them in our choices. There are only two choices available to us as a result (ABR16).

The contradiction in diversity and choice was further seen in the context of Grammar Schools. For example, one respondent stated that:

> having grammar schools in the area restricts our choice (HEN3).

Whereas another blamed the lack of choice on:

> lack of specialisation (streaming), i.e. grammar schools (ABR28).

This highlighted that diversity was, on the one hand, seen as offering choice to some parents, but, on the other, some parents believed that diversity brought with it restrictions on their choice. This was largely because most of the options in the UK education system to offer diversity bring constraints, whether they are based on ability, religion or sex.

Given the concerns of parents in the way they decided which school to send their child, and the limitations to choice that they raised, very few actually stated that they didn't get their first choice of school. Of all the respondents, 90.7% said that their final school was their first choice. This was remarkably high, given the concerns above. But, this did highlight the difference in the way parents perceived their constraints, since when asked for their ideal choice of school, without any restrictions, 16.8% of the respondents mentioned an alternative school to the one they had chosen. This was nearly double the number of parents who said that they did not get their first choice of school. Consequently, parents tended to be realistic in their choice of school and this explained the general level of satisfaction of parents given the considerable constraints and obstacles that they highlighted.

'Parental Choice': A Parents' View

In order to get a parent's view of the education market three elements were covered. The first saw what parents thought about **other** parents' abilities and opportunities in choosing a school. Secondly, parents were asked to consider whether they should have been allowed to choose a school, and finally, saw how parents thought the market should develop and to identify the sort of competition there should be between schools.

Over a third of respondents (38.8%) thought that most parents did not have the ability or opportunity to choose their secondary school. Figure 6.10 shows that this was related to their level of activity in the market place – 'active' parents were the more likely to think that other parents did not have the ability or opportunity to choose a school. Consequently, this showed that the assessment of others' abilities depended largely upon the ability and opportunity afforded the parent concerned. The small minority of respondents who were very 'active' in the market place realised that others could also be more 'active', whereas the majority of the respondents who were relatively 'inactive' saw everyone around them engaging with the market place to the same extent as themselves.

The main reasons given as to why some respondents thought that most parents did not have the ability or opportunity to choose a school were geographical and material resource problems, i.e. based on parents' opportunities rather than ability. As one respondent highlighted:

> the ability to do so is there, but unfortunately the opportunities are few (NOR22).

The greatest constraint mentioned was transport and its related problems. This included access to transport:

> some people wouldn't go into it anyway if there was no possibility of them being able to transport their children to another school (HAR2),

cost of transport,

transport and cost has much to do with where we choose our child to go for education. This is sad, unless you can pay you have no choice (HAR22),

and time and availability of parents,

transport restrictions restrict one parent's availability to work (HAR5).

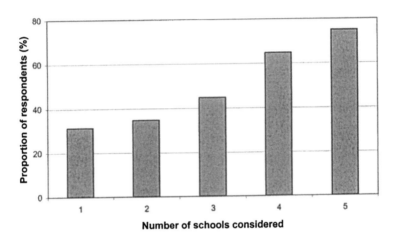

Figure 6.10 Parents who did not think that most people had the ability and opportunity to choose a school, by the number of schools considered during the decision-making process

It was worth noting that rural parents were more likely to comment on distance and travelling restrictions.

Parents also raised problems with the limited amount of choice, but this did not necessarily reflect an individual's ability or opportunity to engage with the market place. However, some parents did go on to say that the amount of choice available had lot to do with where they lived:

Where they live has a huge impact on what is available to them. No one wants his/her child to travel far to school (HAR7).

The problem of 'living in the wrong area' was then linked through to the opportunity parents had in overcoming this since they did not have the:

> ability to afford property which falls into catchment area of chosen school (HEN13).

The admissions procedures was also highlighted as posing problems to parents:

> Schools' admission policies tend to give priority to children from feeder schools (STP7).

But, again, this was largely related to where the parents lived. However, there were other constraints due to admissions procedures:

> Some schools are restricted to some children, i.e. church oriented schools (PEM10).

Even though the majority of the respondents thought that parents did not generally have the opportunity to choose a school, some parents believed that the ability of some parents was also missing. This was typically seen as lacking vision or:

> lack of personal drive (THO10),

and

> lack of interest (HEN12).

Some parents linked this through to material resources:

> Most parents do not have the opportunity because of travelling, lack of finance to send their child to the ideal school. I think the majority [therefore] take the easiest option (HEN11).

As the latter respondent suggested, many of these parents believed that others tended to end up accepting the nearest most convenient school to get to:

> Some people can't be bothered looking at various schools and send their children to the nearest regardless of the school's reputation (HEN4).

Another key constraint on other parents' abilities was because:

> Some people in this area are intimidated when approaching a school for information or where to find out what information is available (ABR20),

or because they

> haven't the confidence to find out [information] or visit schools (THO12).

This lack of confidence was usually linked with the collection of information but could also have been because these parents had a:

> lack of knowledge of [their] rights (POL1).

Very few parents attempted to suggest why some parents lacked the necessary social and cultural capital in choosing a school. However, one respondent argued that:

> I would have thought that only parents with a high standard of education would send their children to a school 'better' than the local one that most of the others would attend (STP34).

Another identified the significance of their community upon this decision by arguing that others might not have had the ability to choose a school because of:

> not wanting to be different and stand out from the crowd (STP33).

This led on to ask if the parents thought that they, and their children, should have been allowed to choose their own school. The overwhelming majority of parents thought that they should. Only nine parents (4.2%) said parents should not be allowed to choose schools, and none of these had children attending the more 'popular' schools. Most of them believed that:

...if all schools were of a good standard and didn't differ so dramatically then parents would not need to choose for any other reason than it being convenient to go to (STP31).

However, the great majority of parents were in favour of choosing a school. Over half of these parents justified their desire for choice because of the different needs of their children:

> Every child is different and local catchment schools do not always suit every family. Secondary education is very important and pupils and parents should be able to choose where this takes place (HEN16).

Once the parents' justifications for choice got beyond the individualism of the child, they fell into three categories; political, social and educational. Politically, parents saw that it was their right to be able to choose a school. This was sometimes because:

> we live in a democratic society and should have choice (HAR11).

But, some parents made the link to free choice as 'consumers' of the state:

> We pay for our children's education and therefore have a right to choose (NOR20).

The social reasons all seemed to suggest that choosing a school meant that one could escape their social environs:

> Kids in the urban estates should not be shoved in to the poor functioning urban schools. Just because they happen to be on the estates the Local Authority expect these kids to go there. Keeping the 'uneducated' uneducated, etc, etc (ABR27),

Giving people choice was seen as giving everyone equal opportunity even though this is meant to be the principle of comprehensive education:

> Everyone should have the same opportunity, why should owning property, or income to afford it, in the right area dictate where your child is sent to secondary school (HEN13).

Educational reasons, other than the individual needs of the children, appeared to be the least important. These educational justifications for choice fell into the desire to improve schools, or to make schools more accountable. Therefore:

> if parents and children had no choice schools would not have to 'try' so hard (POL13).

A considerable number of parents still raised doubts over whether parents should be able to choose schools, even though, in principle, they were in favour. For example, when asked if parents and children should be allowed to choose schools one respondent replied:

> Yes, but within available constraints and resources. Danger is that with over-competition between schools there will be centres of excellence on the one hand and 'sink schools' on the other. Some schools may start introducing their own entrance exams which would be excluding those who are less academically bright (STP43).

As another parent agreed:

> ... I don't believe that this works in practice. It is more likely that the school will choose the child and not the other way around making for a two-tier system (STP29).

The 'natural' progression of choice was, therefore, seen as flawed and, as one parent put it:

> There's no point in raising parents' hopes falsely (HEN10).

As a result of these concerns some parents suggested:

> There needs to be a degree of regulation for the system to work to the benefit of all children (STP22).

However, the suggestions in which 'parental choice' could operate were fairly sketchy and rather limited. For example:

> If the school of your choice is within your catchment area and reasonable access then you should have the right to send your child there (HEN7).

Given these concerns, respondents were finally allowed to suggest ways in which they would like to see schools compete with one another. Table 6.9 shows the results of this. Just over half of the respondents thought that they should be able to choose between schools based on the needs of the children. Whether some parents saw this as a less harsh way of saying that it should be based on ability was unclear. However, 36% of the respondents did say that they thought schools should compete with one another based on their academic performance. Just over a fifth thought that they would like to choose schools based on the specialisation of subjects, while only 10.3% said that the decision should be based on the needs of different communities.

Table 6.9   Parents' views on how schools should compete with one another

How should schools compete?	Proportion of respondents (%)
• Based on what needs the student have (e.g. special educational needs)	50.5
• Based on academic performance (e.g. streaming)	36.0
• Based on specialisation of subjects (e.g. science, drama, or art)	21.0
• Don't know	18.2
• Based on community needs (e.g. culture or religion)	10.3
• No competition	4.7

There were little variations in the characteristics of parents who believed that schools should compete in different ways. For example, there were parents from different geographical areas, with different educational backgrounds and with different levels of household income who thought that competition should be based on the educational needs of the students. The only variation that did exist affected the other two main criteria for competition. It seemed the parents who thought that schools should compete with one another based on academic performance were more likely to be from urban areas than rural areas. Rural parents were more likely to suggest that competition should be based on specialisation of subjects.

Conclusions

This Chapter has attempted to describe and interpret the decision-making process of choosing a secondary school. It has focused on the procedures for choosing a school and the methods employed in choosing a school against the social, economic and spatial constraints that affect the process. On the basis of this analysis a very general model of secondary school decision-making process was developed (Figure 6.11). This shows the key facets that were identified in the process.

The first facet to this model was the presence of a form of order to the process. There was little evidence from the parents that there were stages in the process, but, yet, all three steps identified by Gorard (1996) were recognised. Consequently, there was clear evidence that there were different types of 'consumers' in the decision-making process. First of all, it was clear that parents tended to choose a type of schooling and that from this decision the parents would then consider a subset of schools from which to choose from. These alternative schools generally reflected the size of the competition spaces, identified in Chapter 4, for the areas in which they lived. The parents often made reference to the locality, or locale, when discussing their decision-making strategies. The geographical constraints of proximity and accessibility were also highlighted towards the end of the discussion, in which the relationship between social 'advantage' and the means of transport were an important facet to the choice of schools. The impact of these spatial constraints were most evident in rural areas where parents, with the inclination and social ability to engage with the market place, often did not have the capacity to overcome these spatial constraints, which resulted in a limited set of alternative schools to choose from.

Where parents had chosen 'state' schooling for their child, they chose from schools that reflected the 'hierarchical' nature of the competition spaces. Consequently, it was rare to find schools in the parents' list of alternative schools that were 'below' their nearest school. This suggested that the 'local' school was very important in contextualising the decision-making process. Figure 6.11 illustrates the importance of motives in the process, a facet that has not been critically examined by this or other research. But, clearly, this is a very important force behind the decision-making process and needs to be examined in much greater depth to understand the behaviour of parents when choosing a secondary school.

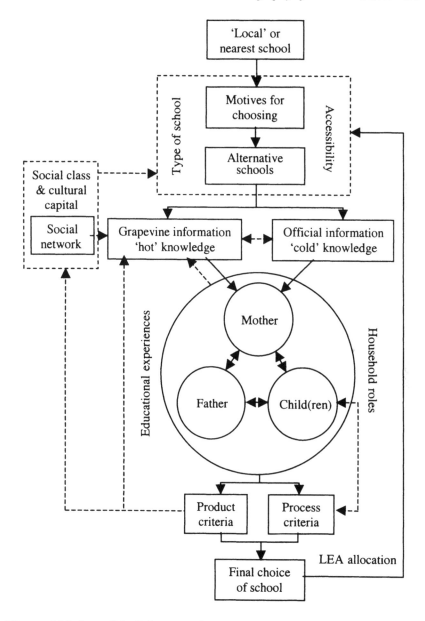

Figure 6.11 A model of the secondary school decision-making process

The research then showed that the next step in the process was the acquisition of information about these alternative schools, from both official, or 'cold', sources, and social grapevine, or 'hot', sources. The extent and importance of these two forms of knowledge were dependent upon their levels of accessibility to parents. It was highlighted that official knowledge was much easier to access and was, therefore, widely used, particularly in urban areas where there were a lot more schools to consider from. However, the grapevine information tended to be of greater importance to parents, but was controlled by the social networks that parents were in. The way these grapevines were used, and the levels of confidence that parents had in using them, was also related to the cultural capital of the parents. Hence, social class was an important facet of the decision-making process, particularly in the learning part of the process.

This research also identified the mother as the key agent in the acquisition of this information. However, this did not necessarily mean that they totally dominated the choice of school. The relationship between the mother, the father and the child often determined who within the household controlled the decision-making process. These relationships were very complex and hence Figure 6.11 only illustrates the importance of these relationships and not how they allocate power and control within the decision-making process. The analysis did suggest that social class was an important element in defining this relationship. Even though mothers generally dominated in the final choice of school, it was argued that they were more influential in working-class families. The relationship between the parents and the child was also related to the social class of the family, with working-class families giving the child a much greater role in the process than middle-class families.

When it came to choosing a school these roles within the household were an important basis on which the decision was made. Alongside parents' own experiences was the importance of product and process criteria (Figure 6.11). Parents often indicated that process criteria were important, but when they were discussing the key reasons for their final choice product criteria clearly dominated. However, the overall importance of process criteria, it was argued, was related to the importance of the child in the decision-making process, after all, process criteria included the 'happiness of the child' and 'a caring environment'. The important product criteria were mainly 'examination performances', reflecting the 'hierarchical' nature of the competition spaces, and the 'reputation' of the school, which, by definition, was based directly on the social networks and the grapevine information that parents participated in.

The final facet illustrated in Figure 6.11 is concerned with the issue of whether the final choice of school was the same school that the LEA allocated a place for the child. According to the growing number of Appeals, presented and discussed in Chapter 3, this would seem to now be an important part of the process. However, in this analysis of the decision-making process, very few parents indicated that they were dissatisfied with their final allocated school. But caution must be added to this, since three-quarters of the respondents said that they faced constraints of some form during the process, and nearly half of them said there was no real choice of schools. This would suggest that the parents in this study rarely optimised the choice that there was, theoretically, open to them. Hence, it would not be wrong to suggest that the education market was still in its infancy and that there is, potentially, greater choice and, therefore, greater competition to come under these reforms. This makes it all the more important that the consequences of market processes in the education system, particularly changes to the social composition of schools, are examined.

7 Social Equity in the 'New' Education Market

Introduction

One of the most under-researched areas of the new education system has been the impact of quasi-market processes on the final allocation of students to schools. This is largely due to the sizeable amount of empirical data necessary to make reasoned observations. This study, and the data employed, provided such an opportunity to examine if there appeared to be any exclusionary processes in the market place that significantly altered the social balance of intakes, and, consequently, would affect social mobility.

Social exclusion in the 'new' education market could potentially occur in three forms:

i.   'Active' participants in the market place.
ii.   Social segregation between schools.
iii. Social polarisation of schools' intakes over time.

It must be noted that these three elements of social change are not unique to the current market conditions being employed in education policy. For example, there has always been a set of parents who were 'active' in choosing schools beyond their allocated catchment school, in particular, those parents engaged in the fee-paying sector of education. There has also been a set of parents who were 'active' in the individual choice of schools within the 'state' sector, and if they did not choose to attend an allocated catchment school then they in effect, chose a school via the housing market, which in turn perpetuated residential segregation. Therefore, if it is acknowledged that there were already 'active' parents in the pre-1988 education market then it can also be safely assumed that there has always been some social segregation of intakes between schools.

Most of the literature on the contemporary market processes in the education system suggests that the market aids the social reproduction of the middle classes. 'The market works as a class strategy by creating a mechanism which can be exploited by the middle classes as a strategy of reproduction in their search of relative advantage, social advancement and mobility' (Ball, 1993, p.17). Of course, the allocation of pupils to schools, that consequently determines the social composition of a school's intake, is of great significance to an equitable society, in particular, the way that schools reproduce social norms and cultural preferences over time (Abercrombie and Warde, 1994). As Halsey (1995) outlines, 'the development of education as an active agent of cultural evolution and social selection is one of the more obvious features of change on British society in the twentieth century' (p. 138).

Therefore, this Chapter will, initially, examine the current debate on social exclusion and attempt to assess the potential significance of exclusionary processes in the current education system in that discussion. Of course, such exclusion can be within the 'new' education market, itself, or exclusion resulting from market processes. The key to these lies in the role education has upon social mobility, and it is this that the discussion will begin. The Chapter will then focus on the three outcomes of potential exclusionary processes – 'active' participation in the market place, social segregation between schools, and social polarisation of school's intake over time – and evaluate their impacts on social equity.

Education and Social Mobility

Longitudinal evidence provided by Halsey, Heath and Ridge (1980), in their work 'Origins and Destinations' of children during the 1970s, showed that there was constant social fluidity between classes throughout post-war Britain. Marshall (1997) went further and suggested that the rate of mobility has increased, and tended to be upwards in the British social hierarchy. However, Marshall also identified that there had not been any greater equality in the *opportunities* for social mobility over time. The power that education can play in the context of social mobility is, traditionally, related to the acquisition of credentials required to enter, and succeed in, the labour market. As Abercrombie and Warde (1994) argued, the most important ideological effect of the British education system is to *position* individuals, depending upon their credentials, within the division of labour. This 'advantage' that could be gained from the education system is not only contained within the realms of employment. Bourdieu and Passeron (1977), in their influential work, 'Reproduction', discovered that the middle-classes in France increasingly capitalised on their cultural assets via the education system. This, the authors believed, also helped to maintain a sense of being disenfranchised by the least 'advantaged' in society. 'Thus, in a society in which the obtaining of social privileges depends more and more closely on possession of academic credentials, the school does not only have the function of ensuring discreet succession to a bourgeois estate which can no longer be transmitted directly and openly. This privileged instrument of the bourgeois sociodicy which confers on the privileged the supreme privilege of not seeing themselves as privileged manages the more easily to convince the disinherited that they owe their scholastic and social destiny to their lack of gifts or merits, because in matters of culture absolute dispossession excludes awareness of being dispossessed' (Bourdieu and Passeron, 1977, p.210).

These discussions of the impact education can have on social mobility suggests there are two perspectives that can be examined. The first is the direct impact that credentials gained from the education process have on the opportunity to enter into, and succeed within, the labour market. The second perspective considers more social and cultural implications that the education system can have in providing the necessary cultural capital and social networks to acquire social 'advantage' that can otherwise be fashioned outside of the labour market. The degree of equality of opportunity for individuals to use both of these facets in increasing their social mobility can be seen in the context of **social exclusion**.

Social Exclusion

Social exclusion has become an enormously popular term in the past 10 years by academics. Even politicians and the media have extensively used it, yet, it faces criticism of being too vague and obsolete by sociologists (see Blanc, 1998). The first application of the term is generally attributed to Rene Lenoir, a member of Jacques Chirac's Government in France in 1974, who identified 10% of the French population as being 'the excluded' (Marsh and Mullins, 1998). This was no more than a precursor to the current use of social exclusion, since it was only used to describe a particular group of individuals. Similar predecessors can also be found in other countries. For example, Townsend (1979) considered a relational concept of poverty to suggest that some members of society were unable to *participate* fully in a 'normal life'. In the USA, the 'excluded' were discussed as a critical response to the growing discourse about an 'underclass' whose *behaviour* was being attributed as the main cause for their poverty (Marsh and Mullins, 1998).

These beginnings led to the concept of social exclusion becoming central to, what has been called, the Europeanisation of social policy-making (Room, 1995). For example, at the beginning of the 1980s social exclusion became the focus of debate in France on a new form of poverty that had emerged from the restructuring of the world economy since the mid-1970s (Silver, 1995). By the late 1980s this had spread to the rest of Europe (Martin, 1996) and resulted in a shift in the European Union (EU) anti-poverty programmes. The first two EU Poverty Programmes (1975-80 and 1986-89) tended to define the poor by their limited material resources, but by the third Poverty Programme (1990-94) this definition was widened to encompass the integration of the 'least privileged' and 'socially excluded'. This change in emphasis continued with the most recent explicit reference to the socially excluded in the Amsterdam Treaty of European Union (1997). Marsh and Mullins (1998) believed that there were two reasons as to why social exclusion became such a key term in the EU social project; firstly, sheer enthusiasm among some of the member states, particularly France, that this concept was useful, and, secondly, there was great difficulty for member states to agree to a definition of 'poverty', resulting in some countries denying that poverty existed in their own countries.

By the 1990s social exclusion had entered almost all discourses on poverty and disadvantage, especially in politics and the media. For example, on election to Government the Prime Minister, Tony Blair, established the Social Exclusion Unit within his Cabinet Office.

The rise in popularity of social exclusion as a concept, particularly because it had entered political rhetoric, prompted the question of whether it is simply 'a product of fashion' (Marsh and Mullins, 1998, p.750). But, in a similar way that Blanc (1998) positively regarded social exclusion, as long as it is carefully defined, using the politically charged 'social exclusion' might have a more direct bearing on policy and public debate.

Since there have been many trajectories for the development of a concept of social exclusion it is not surprising that its definition is fairly contested. Critically, it has been considered as vague, but this is often seen as an important feature, 'the term 'social exclusion' is so evocative, ambiguous, multi-dimensional and expansive that it can be defined in many different ways. Yet the difficulty of defining exclusion and the fact that it is interpreted differently in different contexts at different times can be seen as a theoretical opportunity' (Silver, 1995, p.60). Based on a thorough review of the literature, Silver (1995) suggested that there were three 'idealised' meanings of social exclusion being used and developed: solidarity; specialisation; and monopoly. The many paradigms of exclusion 'are situated in different theoretical perspectives, political ideologies, and national discourses... [and] ...founded on different notions of social integration' (Silver, 1995, p.61).

Generally, the use of social exclusion has arisen through a reaction to the concept of poverty. As already mentioned, within political arenas the definition of poverty faced irreconcilable disagreement. Often poverty was seen as being a very static outcome and largely based on income. Consequently, Room (1994) saw the emergence of social exclusion being represented by a three-fold change in perspective:

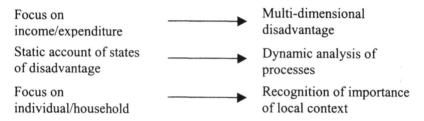

Focus on income/expenditure	⟶	Multi-dimensional disadvantage
Static account of states of disadvantage	⟶	Dynamic analysis of processes
Focus on individual/household	⟶	Recognition of importance of local context

But, as Walker (1997) pointed out, the literature on poverty has gone beyond this rather limited perspective. However, it is argued that poverty and deprivation cannot be more than outcomes of a process that usually entails forms of exclusion in everyday life (Berghman, 1996). This is an important distinction within the context of the 'new' education market. As will be discussed later, there are many ways in which individuals or groups of society can be *excluded* in the market place, but it is difficult to suggest that there is any form of *poverty* in the process. Even as an outcome, it would be difficult to prove that poverty is a result of changes to the organisation of the education system, since the education market continues to guarantee an education to everyone.

This distinction between poverty and exclusion prompted many academics to highlight the special features of social exclusion as a concept. For example, Rodgers (1995) identified five key features that make social exclusion a useful tool in the analysis of social inequality; it provides a multi-dimensional and multi-disciplinary view of inequality, it focuses on processes of inequality, and focuses on the social actors and agents within those processes, it examines the impact of inequality at many levels, and can use the terms of inclusion as a strategy for reducing inequalities.

It is generally considered that there are two ways in which members of society are at risk from being excluded. The first, and most accepted, is the risk of being barred from the labour market. However, as Rodgers (1995) suggested, there was also the risk of being excluded, not just from the labour market, but also from regular work. The second main area in which one can be excluded is, as Martin (1996, p.382) stated, 'the risk of seeing one's network of social relations and primary integration broken up because basic social links ... disintegrate'. This second area of exclusion is often associated with social citizenship (Somerville, 1998) and the resulting denial of social citizenship status to particular groups. Gore (1995) elaborated on this, and proposed that both citizenship and social integration underlie any notion of social exclusion. As a result, Shucksmith and Chapman (1998, p.230) provided a useful working definition of this second main area of social exclusion as 'a multi-dimensional, dynamic concept, which refers to a breakdown or malfunctioning of the major societal systems that should guarantee the social integration of the individual or household'.

Inequity and Social Exclusion in Education

The social composition of school intakes has been a key policy target during the era of comprehensive education. As Ball (1986) suggested, there has never been a qualified definition of comprehensive education, but one of the key aims was for an 'integrative approach' to the education system. The Labour Party actively sought this during the 1950s and 1960s, when it was thought that having socially mixed intakes would produce greater tolerance and social harmony to society as a whole. Consequently, a change in policy away from this integrative approach falls into the terms of inclusion and exclusion.

So, the question that needs to be asked is whether recent changes to the education system in England and Wales have exacerbated the inequality of opportunities for social mobility. Social exclusion in the 'new' education market will have implications on the degree of *risk* members of society face in two key areas of social inequality. Firstly, if the market excludes some parents from being able to send their children to the schools where their children are most likely to achieve the greatest levels of their ability, then this will severely have an impact on their chances in the labour market. This, consequently, goes against the principle of meritocracy. Secondly, if the market excludes some people from being able to participate in the market place, then there is the possibility that the socio-economic composition of intakes will become more segregated. This would, therefore, reduce the effects of having a socially integrated intake and reinforce social networks, and cultural norms and preferences, along traditional social cleavages.

There has been considerable debate about the ways in which the education market may be socially exclusive. The most accepted cause for exclusion in the education system generally, lies in the degree of **cultural capital** that parents have (Bourdieu and Passeron, 1977). Brown (1995) went further and added that it is necessary to have market power, resources rather than influence, to succeed in the 'new' education market. 'The increased opportunity for the middle classes to exert the full weight of their market power in the competition for credentials will ensure that they will seek to dominate access to elite institutions at each stage of the education process – from the cradle to graduation and beyond' (Brown, 1995, p.44). With greater cultural capital or market power it is assumed that some parents will take greater advantage of the education market than others by getting their children into particular schools, thereby maintaining, and perhaps increasing, their relative social advantage. But, in order to examine the full implications of the education market on social mobility, the depth of study undertaken, for example, by Halsey *et al.* (1980) and Goldthorpe (1980 and 1987) throughout the 1970s and 1980s, would be required, particularly to identify the effects on an individual's success in the labour market post-schooling.

It is possible to start examining the consequences of the education market at an *initial* level in order to answer the following questions, based on a similar approach by Willms and Echols (1992) in Scotland:

- Do parents who engage with the market place differ in their socio-economic characteristics from those that accept their 'local' school?
- Does the education market serve to reinforce or exacerbate the segregation of particular social networks?
- Which parents choose schools that will improve their child's predicted level of performance on examination?[1]

These important questions about the impact of the 'new' education market on social exclusion of members of society will be considered using the following three indicators of exclusion:

i.   'Active' participants in the market place.
ii.   Social segregation between schools.
iii.   Social polarisation of schools' intakes over time.

'Active' Parents in the Market Place

Research that has tried to distinguish between the 'active' and 'inactive' ('alert' and 'inert' in Willms and Echols, 1992) parents in the market place has been far from conclusive, largely because it has been difficult to define the levels of engagement with the market. However, earlier research in

Scotland managed to avoid having to distinguish between 'active' and 'inactive' parents in the market place, by simply identifying the socio-economic characteristics of those parents who placed official requests for a choice of school other than their designated school (Raab and Adler 1988; Adler *et al.*, 1989). In these studies both middle-class, and working-class families, were found to engage with school choice. Similarly, the PASCI (Parent and School Choice Initiative) survey, found that approximately three-quarters of the working-class families in their sample thought that they had a choice of school (Glatter and Woods, 1994), which contrasted quite significantly with the middle-class families, whom only about half believed they had a real choice. However, as Glatter and Woods (1994) argued, these variations were based on a perceived level of choice rather than identifying different levels of actual engagement with the market.

Another attempt to differentiate between 'active' and 'inactive' parents was by Carrol and Walford (1997), who distinguished between 'active' and 'passive' parents using a fairly extensive set of criteria for different *types* of education 'consumer'. Based on this rationale, Carrol and Walford (1997) found that parents from high social classes were more likely to be 'active' in the market place than those from low social classes, yet, they still found that there was a significant minority of working-class parents who were also 'active' in the market place. The criterion that Carrol & Walford (1997) used to identify the 'active' and 'inactive' parents was relatively extensive and well considered. However, some of the criteria were specific to the case study area and could be criticised for not reflecting the levels of engagement with the market. For example, one of the criteria for the 'passive' type of consumer was the recognition by parents that the child would achieve the same results irrespective of the school attended. This rather intuitive observation by parents need not necessarily mean that they will not, and indeed, would not, engage in the rhetoric and activity of school choice.

Unfortunately, much of the remaining research on levels of activity in the market place often seem driven by theoretical notions of inequality in empowerment and ability to engage with the market, particularly, the writings of Bourdieu (1986) relating cultural capital to social exclusion, and the discussions by Ball (1997) on the recent education policy shifts in the UK. Consequently, it is difficult to determine from the literature how the degree of engagement with the market was defined. For example, Conway (1997) discussed the way in which the working-classes failed to engage with the market place due to class-related inequalities in power and knowledge, but, beyond the rhetoric of choice, Conway really only defined the level of engagement in which these conclusions were based by the *distance* parents were willing to send their children. 'The more middle class parents would consider schools out of their immediate locale if they were deemed to have properties which would increase the lifechance prospects of their children' (Conway, 1997, 4.3).

Similarly, Stillman (1990) in a survey of 1,792 parents found that all social classes participated in school choice to some degree, but that the middle-classes took greater advantage of having the opportunity to choose the school for their child. Once again, the *distance* parents were willing to send their child was used as giving 'advantage' to these parents. 'The longer they were in full-time education and the higher their job classification the more information they used and the more likely they went to choose a more distant school' (Stillman, 1990, p.101).

In order to determine the significance of these findings this study has used several different ways of defining the levels of engagement with the market place, largely as a result of the availability of data at different scales. Table 7.1 provides a summary of three ways in which the level of activity in the market place was determined. The *distance* definition simply distinguishes between the 'inactive' parents, who sent their children to their nearest school, and the 'active' parents, who sent their children to an alternative to their nearest school. This definition of activity used both a small sample, based on household level data, and a larger sample, using 34,178 pupil postcodes.

Table 7.1   Working definitions of levels of engagement in the market place

Definition	Active	Inactive	Small or large sample*
Distance	Local	Non-local	Both
Extent of choice	Continuum based on the number of schools considered: 7 ◄――――► 0		Small
Probability	Continuum based on the probability of attending a school: 0.0 ◄―――► 1.0		Large

*   Small sample refers to the responses of 214 parents. The large sample refers to the use of 34,178 pupil postcodes.

The second definition given in Table 7.1, *Extent of choice*, used the household level questionnaires, which asked parents how many schools they considered during the school choice process (see Chapter 6 for further elaboration on this). This assumed that the number of schools considered during the process related directly to the level of engagement with the market, i.e. the greater the number of schools the higher the level of activity.

The third definition for the level of activity in the market was based on the *probability* of a child attending a particular school. This probability was calculated in ArcInfo (GIS), using the Interaction function, and applied a distance decay factor against the spatial distribution of competing schools for each individual household. This gave the probability of a child, from each household, to attend any school within their Local Authority. How this was calculated meant that the individual probabilities were based on the *relative* location of a school to all its competitors, which also meant that the sum of these probabilities, for each child, was equal to one. Consequently, it could be argued that, rather than just using a measure of distance between schools to identify those parents who appeared to be more 'active' than others, this accounted for *relative* variations in the spatial distribution of schools within each LEA. The results meant that the probability of each child attending their present school would give some more accurate indication of 'spatial effort', which in turn represented the level of activity by families in the market place.

The socio-economic data used in this Chapter uses geodemographic profiles and their associated variables for households, based on their postcodes. There are a number of potential problems with using socio-economic data generated from postcodes (Wilson and Elliot, 1987; Raper *et al.*, 1992), however, this is alleviated by the use of a purposely constructed geodemographics package (GB Profiler '91) that combines many elements from the 1991 UK Census into aggregated and standardised key variables (Openshaw and Wymer, 1995).[2]

The 'Distance' Definition of Market Activity

From Figure 7.1 it can be seen that, using the *Distance* definition of 'locals' and 'non-locals', there appeared to be little difference in the socio-economic character of the two sets of parents. Household level data analysed in Chapter 6 supported the conclusion that there was little difference between the respondents who sent their children to the 'local' catchment school from those who sent their children to an alternative school. This household level data was based on a relatively small sample

(214 respondents). On a larger scale, using the postcode data of 34,178 pupils, there was still little difference across all the study Local Authorities in the composition of the 'local' and 'non-local' parents (Figure 7.1).

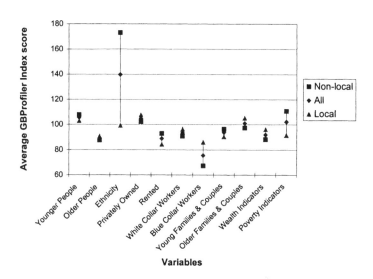

Figure 7.1 Difference in socio-economic characteristics of 'locals' and 'non-locals' across eight LEAs

However, there were three variables that appeared to indicate some differences between 'locals' and 'non-locals': ethnicity, blue-collar workers, and poverty indicators. The most obvious feature was that 'non-locals', or the more 'active' in the market place, tended to have high scores of ethnicity. It would have been more accurate to reverse this and say that pupils in the study with high ethnicity scores tended to be more 'active' in the market place than those from neighbourhoods with low ethnicity scores. The significance of ethnic minorities in the market place has been identified elsewhere. For example, Tomlinson (1997) discovered that middle-class minority groups, particularly from Indian and East African Asian ethnic origins, were very 'active' in the market place, particularly for Muslim girls (Basit, 1995). Similarly, the Audit Commission (1996), in their report, 'Trading Places – the supply and allocation of school places', focussed on Bordesley Green Girls School in Birmingham, which, because it was a single-sex school, attracted Muslim girls from a very wide area. But, as Tomlinson (1997) has explained, there is still an enormous class cleavage within ethnic minority groups when studying their levels of engagement with the market. Figure 7.1 also shows that, perhaps surprisingly, the most 'active' parents within the market, under this definition, had a higher aggregate poverty indicators index score. This was largely explained by the underlying variation in market activity caused by the spatial constraints seen in the urban and rural context. The third variable that identified socio-economic differences between 'locals' and 'non-locals' was that of blue-collar workers (Figure 7.1). This suggested that the more 'inactive' parents, or 'locals', in the market place tended to be blue-collar workers. Once again this reflected the urban-rural contrast between Local Authorities, because in rural areas large farming communities are classified as blue-collar workers, and since these areas faced significant spatial constraints to choice it was not surprising that, altogether, 'locals' had a high blue-collar workers index score.

Discussing these three variables has shown that there was little real evidence of socio-economic differences, using the *distance* definition of market engagement, across the study areas between the more 'active' and 'inactive' parents in the market place. Taking each LEA in turn really only emphasised this point.

The 'Extent of Choice' Definition of Market Activity

Of the three definitions of market activity being examined here, probably the most accurate identifier of the level of engagement with the market was the *extent of choice*, or number of schools, considered by the parents during the decision-making process. The features of the 'active' and 'inactive' parents, as defined by the *extent of choice*, were discussed in more length in Chapter 6. But, in summary, based on the socio-economic characteristics given by the respondents in the questionnaire, it was argued that the more 'active' choosers were over-represented by parents who were educated privately and had been to university. These differences were more obvious than when comparing 'locals' and 'non-locals', but there were still parents from lower social classes who were engaged to a high degree with the market in the choice process. This was highlighted by the aggregate GB Profiler '91 index scores, given in Figure 7.2, according to the number of schools that the respondents indicated in the survey they chose from. For the majority of the variables there was little variation, though a number illustrated differences between parents who only considered, say, one to four schools, and those that considered five or more schools. For example, both the levels of poverty indicators and rented accommodation variables dropped with the more 'active' of these two sets of parents. Conversely, the levels of white-collar workers, wealth indicators, and privately-owned accommodation variables all rose with the greater level of market activity. This suggests that there were differences in the socio-economic characteristics of parents who were *very* 'active' in the market and those that were relatively 'inactive'. However, it must be pointed out that 97% of the respondents considered four or fewer schools, and, according to Figure 7.2, there was little difference between the parents who considered one, two, three or four schools.

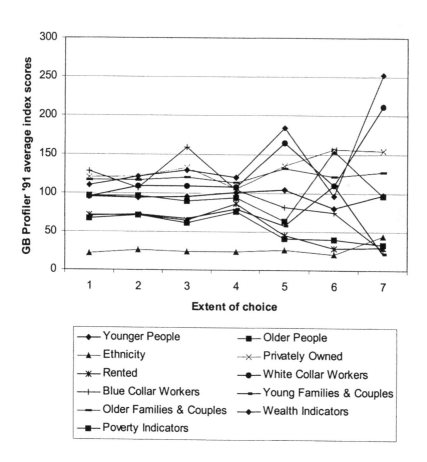

Figure 7.2 Socio-economic characteristics of parents by their extent of choice

The 'Probability' Definition of Market Activity

The final definition of market activity of parents being considered here was the *probability* of the children attending a particular school (Table 7.1). Based on the concept of gravity models, this reflected the 'effort' to which parents had gone in order to overcome the 'friction' of distance in sending their children to particular schools. Table 7.2 provides a summary of the average probability of parents, from five Local Authorities, calculated for the school they actually sent their children to. These probabilities were between 1 and 0 – the lower the value the less likely that the child would have attended a school based on their distance to that school and the relative distances to other schools within the LEA.

Table 7.2 **Proportion of intake by probabilities of attending present school**

Proportion of intakes (%)	*Probability Clusters* Market activity				
	0.0-0.2	0.2-0.4	0.4-0.6	0.6-0.8	0.8-1.0
Large metropolitan borough (West Midlands)	72.5	13.8	6.7	4.3	2.7
Small metropolitan borough (West Midlands)	70.5	24.1	4.3	1.1	0.1
Metropolitan borough (Greater Manchester)	51.0	18.5	12.9	8.8	8.8
County (Eastern)	24.7	23.0	14.7	12.9	24.7
County (West Midlands)	57.2	29.5	8.7	3.9	0.7

Table 7.2 shows, for example, that parents in the Eastern county LEA generally chose schools nearer to them, and relative to other school locations, because of the greater distance constraints of choice in rural areas. However, this was in stark contrast with the other rural Local Authority in the West Midlands, where many parents sent their children to schools that, according to the gravity model, had low probabilities of them

choosing. The urban large metropolitan borough in the West Midlands had an even lower average set of probability scores, in keeping with the greater market activity and competition identified in Chapters 3 and 4.

The average GB Profiler '91 index scores for each probability cluster are given in Table 7.3. For the majority of the variables there were significant differences between the average index score of the most 'active' (p = 0.0-0.2), and the least 'active' (p = 0.8-1.0), sets of parents, and in each case the most 'active' had the greatest social 'advantage'. This compared the most *extreme* levels of 'consumer' activity in a similar way to the *extent of choice* definitions of activity. However, in this example, the number of pupils within each probability cluster was more evenly spread. This pattern was reinforced where there was a direct relationship with the index score and the level of activity in the market place in a number of cases. For example, in the small metropolitan borough (West Midlands), the county LEA (West Midlands) and the metropolitan borough (Greater Manchester), the higher the levels of ethnicity, the higher the levels of market activity.

In the large West Midlands metropolitan borough there were two peaks of high ethnicity, one for the most 'active', and one for the least 'active', in the market place. This reflected the complexity of ethnicity and school choice, and the potential social cleavage that polarised ethnic minorities, as either, extremely 'active', or extremely 'inactive', as suggested earlier.

The other notable feature highlighted in Table 7.3 was for the two variables, wealth indicators and poverty indicators. As already stated, the aggregate values for the most, and the least, 'active', reflected the social 'advantage' that 'active' parents had in the market place. However, the interesting observation of these two variables was that there was only a direct relationship between the levels of activity and social 'advantage' for the two rural county LEAs. But for the majority of all parents, say, with probability scores between 0 and 0.8, there was little variation in their relative social 'advantage'. This suggested that with greater market activity and competition overall the variation in the socio-economic characteristics of differing levels of engagement with the market was markedly reduced. The exception to this was the small number of parents, who were at best described as being relatively socially 'disadvantaged', who were yet to participate in the market to the same extent as the majority of parents in each Local Authority.

Table 7.3 Average GB Profiler '91 index scores by probability cluster

	Probability Clusters	Proportion of intake (%)	Younger People	Older People	Ethnicity	Privately Owned
Large Metropolitan Borough (West Midlands)	0.0-0.2	72.5	115.4	84.6	292.7	98.3
	0.2-0.4	13.8	113.5	85.1	262.9	99.5
	0.4-0.6	6.7	116.0	83.7	285.7	96.3
	0.6-0.8	4.3	116.4	85.0	306.0	90.8
	0.8-1.0	2.7	114.4	85.4	330.8	92.8
Small Metropolitan Borough (West Midlands)	0.0-0.2	70.5	106.3	87.9	92.3	96.0
	0.2-0.4	24.1	103.3	88.8	62.7	94.7
	0.4-0.6	4.3	97.2	92.0	41.2	113.8
	0.6-0.8	1.1	101.9	98.6	27.3	107.2
	0.8-1.0	0.1	91.0	120.2	31.3	62.5
Metropolitan Borough (Greater Manchester)	0.0-0.2	51.0	104.8	81.3	30.1	111.9
	0.2-0.4	18.5	106.0	76.3	29.6	116.2
	0.4-0.6	12.9	105.1	83.8	29.3	102.9
	0.6-0.8	8.8	107.8	80.7	29.7	100.0
	0.8-1.0	8.8	104.7	85.2	29.0	104.8
County (Eastern)	0.0-0.2	24.7	93.7	102.5	24.6	121.4
	0.2-0.4	23.0	94.9	97.4	25.9	123.0
	0.4-0.6	14.7	95.9	94.8	27.2	125.6
	0.6-0.8	12.9	94.5	105.6	27.7	116.7
	0.8-1.0	24.7	96.4	100.7	27.9	113.3
County (West Midlands)	0.0-0.2	57.2	99.8	86.3	36.4	120.9
	0.2-0.4	29.5	98.7	88.9	34.9	121.3
	0.4-0.6	8.7	99.4	87.9	32.2	116.8
	0.6-0.8	3.9	100.4	84.5	27.6	119.2
	0.8-1.0	0.7	98.6	84.1	24.8	118.7

Table 7.3 (Continued)

		Rented	White Collar Workers	Blue Collar Workers	Young Families & Couples	Older Families & Couples	Wealth Indicators	Poverty Indicators
Large Metropolitan Borough (West Midlands)		101.5	76.7	52.9	102.1	88.3	74.5	134.8
		97.1	75.7	54.4	100.5	90.6	70.6	127.8
		100.9	73.1	54.5	101.8	89.8	71.9	133.7
		107.7	68.2	53.1	103.8	86.0	65.6	139.5
		104.9	70.2	53.2	93.7	85.5	65.6	131.9
Small Metropolitan Borough (West Midlands)		85.3	109.3	177.1	78.0	117.8	130.8	68.3
		75.1	115.2	130.3	75.8	118.4	132.0	66.1
		66.2	113.3	105.9	82.6	116.6	120.6	68.6
		78.4	98.1	101.7	88.4	111.7	102.9	77.1
		76.5	97.5	91.4	87.4	111.1	98.7	76.6
Metropolitan Borough (Greater Manchester)		89.1	80.2	64.0	98.3	101.1	75.2	102.5
		86.8	77.1	64.6	91.3	103.0	68.8	101.1
		68.7	90.0	68.2	77.0	111.5	87.5	80.2
		79.6	72.2	64.6	93.7	105.9	68.8	87.3
		116.3	43.8	56.7	89.3	97.0	32.8	136.9
County (Eastern)		68.2	120.8	107.8	81.9	116.2	125.9	63.7
		61.7	112.5	75.3	80.9	114.0	109.0	64.3
		62.0	114.6	72.9	81.7	114.2	107.3	61.2
		60.9	110.2	80.9	82.6	114.9	110.4	59.7
		64.7	112.8	86.3	83.1	114.8	109.4	59.1
County (West Midlands)		66.4	93.0	70.2	95.4	107.8	80.0	77.7
		61.7	104.4	68.7	94.8	110.2	90.7	74.0
		76.2	89.0	67.7	97.6	105.6	76.7	83.0
		78.5	89.0	66.7	99.4	104.7	76.9	87.7
		74.5	82.0	66.6	95.4	104.9	69.7	84.3

Levels of Market Activity

Having considered three definitions of market activity it became clear that members of all social classes participated, *to some extent*, in the education market. In terms of the number of schools that parents considered during the process then the most 'active' did appear to have greater relative levels of social 'advantage' compared with the rest. Entry to the nearest school and the resulting probabilities of attending a particular school revealed a similar level of market activity. However, this did provide further insights into the importance of time in the development of the market place. Local Authorities that had little market activity, overall, were the same LEAs that tended to have a direct relationship between the levels of engagement for individual parents and their relative social 'advantage'. Then, in Local Authorities with greater market activity overall, the socio-economic differences between the 'active' and the 'inactive' dissipated with greater participation of **all** social classes in the market place. However, the few parents who did not appear to participate to any great extent, whichever definition was used, were typically characterised by their lack of social 'advantage'.

This distinction between levels of market activity and greater social *inclusion* in the education market has been identified elsewhere (Glatter and Woods, 1994; Waslander and Thrupp, 1995). But, the attempts to explain the different levels of engagement, and their impact on 'advantage' in the market, have been limited. Probably the most useful analysis and discussion has been by Ball *et al.* (1995), and their concept of **circuits of schooling**. Based on research in three Local Authorities in London the authors believed that there were four circuits of schooling, each of which related differently to choice, class and, significantly, space:

i.   Local, community, comprehensive schools.
         Intake from immediate locality, highly localised reputations and have policies and structures which provide given them a comprehensive identity.
ii.   Cosmopolitan, high profile, elite, maintained schools.
         Intake from outside immediate locality, reputations that exist beyond locality, selective.
iii.   'Local' system of independent day schools.
         Compete with maintained sector and private sector
iv.   Parallel circuit of Catholic schools.
         Own hierarchy, pattern of competition and spatial structure.

Despite the elegance of Ball *et al.*'s classification there appeared to be two problems with the idea of using space, and more accurately, distance, as a measure of 'advantage' in the education market place. First, the only time that any link was made between a given level of engagement with the market and social inequality was when they were associated with a discussion on the 'ability' to consume within the market (Ball *et al.*, 1996). This produced three ideal types of school choosers: the privileged/skilled; the semi-skilled; and the disconnected. However, this classification only identified the 'advantage' and 'disadvantage' of choosing schools by the way in which the parents went about making a choice of school. It did not really associate the circuits of schooling with advantageous, or disadvantageous, choices of schools.

The second problem with Ball *et al.*'s study was that it appeared that there was no limit to the spatial extent of parents' decision-making, which kept the circuits of schooling distinct from one another. However, as revealed in Chapter 6, nearly three-quarters of the respondents in the present survey believed there were a distance and a time limit to their choice of school, and, of great importance, this appeared to be irrespective of social class. Using the actual distances, measured in ArcInfo (GIS), between the homes of pupils and the schools they attended, there emerged a **series** of critical distances that they travelled. Figure 7.3 shows these critical distances, or thresholds, for seven Local Authorities in the study.

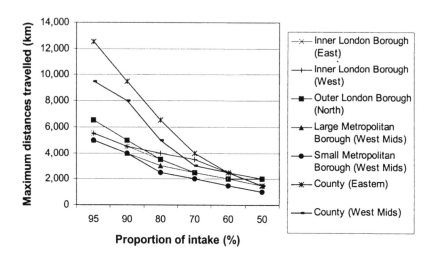

Figure 7.3   Critical distances travelled to school, by LEA 1995

This Figure shows that, even with taking into consideration the relative distances between schools within an Authority, the distances travelled tended to reach a threshold for a significant majority of parents. So, for example, 90% of urban parents did not send their children further than 5km away from their home, and 80% of them travelled no further than 4km to school. However, again this emphasised that a small minority of parents appeared to be significantly more 'active' than their counterparts.

These two problems highlighted the obvious difficulty in associating some notion of market activity with social exclusion, and, therefore, limited any observations from being conclusive. This was compounded further if one follows the argument of Reay and Ball (1997) that middle class norms were being used to interpret working-class school choices. Consequently, it could be argued that the sense of locality in school choice, and the assumption that this 'advantages' some parents over others, might have been exaggerated by a middle-class understanding of school choice success.

In summary, this section has tended to be critical of any conclusions emanating from the variations in the level of market engagement from three perspectives. The first was in the way that the degree of market activity was measured or identified. The second perspective was that the social 'advantage' of those who participated to a greater extent in the market place might have been exaggerated. This study has shown that, for the majority of parents, there were little social class differences between their relative levels of engagement with the market. Given this, however, caution must remain, since the findings have also shown that for a relatively small number of parents their fate in the market place was linked to their social class. There were a number of parents who were clearly more 'active' in the market than the majority, and a number who also appeared to be *very* 'inactive' in areas where everyone around them was 'active'. The contrast in the social status between these two sets of consumers was quite marked.

The third critique emerging from this discussion, irrespective of whether there were links between the social class of parents and their ability to engage with the market, was it possible to say that this **market 'advantage'** could be transferred to **social 'advantage'**? One way of attempting to answer this question is to look at the changes in the composition of school intakes.

Social Segregation Between Schools

The analysis in Chapter 4 showed that the relative success of a school within individual *competition spaces* was typically associated with the relative examination performance of their intakes. Therefore, it was possible to say that the 'attractive' schools in the market place generally had higher examination performances than others. Consequently, if being more 'active' in the market meant that parents were more likely to seek out and get a place in such schools, then they might have been 'advantaged' because their children were in a school which would raise the likelihood of them doing well in their own examination performance (Willms and Echols, 1992).

However, in order to see if the higher social classes were really seeking out the best performing schools it was necessary to focus on the social composition of intakes and the potential social segregation between schools.

Social segregation of school intakes has always been around (see Chapter 3 and Ball *et al.*, 1996), and refers to the allocation of pupils to schools by their social class. Even the use of catchment areas during the era of comprehensive education simply reflected residential segregation (Spring, 1982). Indeed, one of the social equity arguments in support of the market in education was it allowed parents to get out of this 'trap' of having to attend the school associated with their residential locale (see Coons and Sugarman, 1978; Chubb and Moe 1990). Chubb and Moe (1992, p.46) gave the following response to the supposed Labour Party opposition to choice in education, 'the great irony is that the common man is the real victim of the traditional system. People with money do quite well. They can move to the suburbs in search of good schools or pay for private schools. But most ordinary people in the inner cities, especially the poor and minorities, are stuck. The system provides them with lousy schools, and they have nowhere to go.'

It might not, therefore, be surprising that Lee *et al.* (1994) found that, from their research in the USA, socially disadvantaged adults were the strongest supporters of school choice. Also, Menahem *et al.* (1993) attempted to link the effects of school choice with residential segregation and similarly found support for the social equity argument of markets in education. Based on reforms in Israel, Menahem *et al.* asked parents if they would have moved to particular areas in order to get the school that they eventually chose under a system of open choice. Their research discovered that allowing parents to choose a school was actually moderating processes of residential segregation. Both Lee *et al.* (1994) and Menahem *et al.*

(1993) questioned the findings of their own research and would not say that this evidence in support for social equity through school choice was conclusive. Realistically, both sets of research were empirically limited, but it was not surprising that they had reservations, given that the majority of documented evidence purports the social inequalities of the market in education.

Much of the research on the market in education believed that open enrolment would eventually exaggerate the existing social segregation of schools. Ball (1993) believed that the ideology of the market was the mechanism that drove class reproduction, and that this worked in three interrelated ways:

i.   It assumes that the skills and predisposition to choice, and cultural capital, which may be invested in choice, are generalised.
ii.   It legitimates differences in these by labelling non-choosers and poor choosers as 'bad parents'.
iii.   The education market doubly disadvantages the 'poor choosers' by linking the distribution of resources to the distribution of choices.

This 'negative' outcome of the market has also appeared to occur in other European countries. For example, Ambler (1994, p.372) used evidence from France, The Netherlands and Great Britain in order to show that 'the primary negative effect of school choice is its natural tendency to increase the educational gap between the privileged and the underprivileged' by increasing the social segregation between schools. Similarly, Waslander and Thrupp (1995) found segregating tendencies for schools in New Zealand.

However, most research in the UK on the new education market has tended to be *process-based*, consequently, there have been few empirical studies on the actual changes to the composition of intakes as a result of these processes. One key exception to this has been the work of Gorard (1997 and 2000), but these findings are discussed later. It is the outcomes of the market processes that provide a more complete understanding of whether there is greater social inequality and social exclusion due to the reforms.

Ideally, in order to analyse the social segregation of schools, the socio-economic characteristics of year-on-year intakes are needed. Unfortunately, this study only had one year's intake and had to determine social segregation by the means of open preference, with and without, the effects of the market for just that one year. This was possible by taking the socio-economic characteristics from two different versions of the intake of a school:

i.   **All local** – based on thiessen polygons as a surrogate for the traditional catchment area.

ii. **Actual** – based on the intake derived under the new market conditions.

These two versions of a school's intake reflected, quite straightforwardly, the 'before' and 'after' of the market process. Composing the aggregate socio-economic characteristics of these two possible intakes for each school gave some indication of whether the schools had become more socially segregated.

To see if there had been an overall change in the degree of social segregation between schools in each Local Authority, the variance across the LEA was calculated for the 'before' and 'after' school intakes. The difference between the standard deviations of these two sets of school intakes was then calculated, which indicated whether the schools across each Local Authority were becoming more, or less, similar in their social composition. Table 7.4 shows that there were very few occurrences, within each of the eight Local Authorities in this study, where these variations between schools had actually increased. In other words, the distribution of social classes across schools in these authorities had actually converged. In nearly every Local Authority, and for every GB Profiler '91 variable, the standard deviation between each school intake had fallen, suggesting that the overall social segregation *between* schools was actually less than it would have been if everyone had gone to their nearest school.

It was also worth noting that the Local Authorities with the greatest reduction in the distribution of social classes were those with the greatest market activity overall. The reduction in the variance between school intakes was generally greater for the Eastern Inner London borough, the large West Midlands metropolitan borough and the metropolitan borough in Greater Manchester than it was for the small West Midlands metropolitan borough, the West Midlands county and the Eastern county.

Table 7.4   Overall segregation of schools within each local authority

LEAs	[= (standard deviation of school intakes 'after') – (standard deviation of school intakes 'before')]				
	Younger People	*Older People*	*Ethnicity*	*Privately Owned*	*Rented*
Inner London borough (East)	-3.24	-5.06	-34.07	-10.92	-12.11
Large metropolitan borough (West Midlands)	-2.06	-1.98	-26.08	-8.12	-8.12
Small metropolitan borough (West Midlands)	-0.30	-3.57	-20.78	-0.24	-1.54
Metropolitan borough (Greater Manchester)	-2.14	-6.97	-3.01	-0.34	-2.26
County (Eastern)	-0.23	-1.51	-0.26	-1.27	-2.05
County (West Midlands)	-0.74	-1.99	-1.88	-2.48	-5.45

This analysis showed that, generally, the overall variations in the social composition of schools across each Local Authority were less with open enrolment, and that they were lower in Local Authorities with the greatest market activity and competition. Given the general leaning of the literature and other research it was surprising that there appeared to be little 'between-school' segregation as a whole. However, research by Gorard (1998) in Wales has also shown that there was relatively little 'between-school' segregation, and that, over time, it was falling. This research used free school meals as a surrogate measure for low social class over the period 1990/91 to 1996/97, and was then used to create a segregation index (Gorard, 1998; Taylor *et al.*, 2000), which suggested that, for the whole of Wales, there was little change to the overall 'between-school' segregation. Local Authorities in South Wales, which, incidentally, were more 'developed' market places than other authorities in Wales, actually saw a

Table 7.4   (Continued)

[= (standard deviation of school intakes 'after') – (standard deviation of school intakes 'before')]					
White-collar Workers	*Blue-collar Workers*	*Young Families*	*Older Families*	*Wealth Indicators*	*Poverty Indicators*
-9.31	-9.62	-9.78	-5.03	-8.99	-12.43
-6.32	-2.53	-4.09	-2.48	-5.70	-7.54
-0.91	-2.20	-1.76	-0.87	-1.24	-1.42
-6.11	-0.37	-7.27	-1.76	-4.32	-4.53
-0.35	8.31	-1.42	-0.37	0.96	0.21
-2.18	-22.73	-3.84	-0.94	0.01	-3.48

reduction in the segregation index from 24% in 1990/91 to 21% in 1996/97 (Gorard and Fitz, 1998). This indicated that 'between-school' segregation had fallen, and that the level of overall market activity and competition was possibly driving this down. Quite simply this meant, 'popular schools … are increasing their proportion of children from economically disadvantaged families' (Gorard, 1998, p.254). Gorard and Fitz (2000) have since applied the segregation index to secondary schools in England and revealed that apart from an initial rise in segregation between 1989 and 1991 there has been a general decline in socio-economic segregation between schools. Their findings were very much in keeping with the results of this study.

In summary, the analysis has found little evidence of significant increases in the existing social segregation of intakes, but it is argued that this does not necessarily mean that, within each Local Authority, the intakes of **every** school were becoming more homogenous. Indeed, were there schools that had intakes that were becoming more socially distinct from other schools in the Local Authority?

The discussion in the previous section suggested that, in general, there was little difference in the level of activity between the majority of households. But, it also identified that there was a minority of the population in which social exclusion and inclusion were having effects on the process of school choice.

In order to find out if there were some schools becoming more socially segregated from the rest, it was necessary to focus on the intakes of each individual school and the social polarisation that arose.

Social Polarisation of School Intakes

The previous section identified if there was 'between-school' segregation across each Local Authority in this study. The results showed that, generally, this segregation was less with open enrolment than it would have been if thiessen catchments were being used to allocate students to schools. However, this LEA-level analysis on the outcomes of the market upon school intakes might have hidden some very important features. Consequently, this section examines each individual school, and, rather than see if their intakes were becoming similar as a whole, it will identify whether their intakes were becoming more polarised in relation to the socio-economic characteristics of all the students from their respective LEAs. Using a measure of social polarisation for each school allowed further analysis of the underlying reasons as to why the socio-economic characteristics of intakes were changing in particular ways.

Calculating Social Polarisation of School Intakes

In order to identify if a school had become socially polarised, the change in the socio-economic characteristics of the intake was compared with some relative point in which all the schools were measured. To do this the intake profiles used in Chapter 5, which identified the 'before' ('all catchment' intake profile) and the 'after' ('actual' intake profile) situations, need to be reintroduced. This analysis compared the change between the 'all catchment' and the 'actual' intake profiles with the average socio-economic characteristics of the relevant Local Authority. This was undertaken for each of the 11 GB Profiler '91 variables, and measured these comparisons against the mean of each Local Authority to see if the socio-economic composition of the intakes was polarising away from the LEA average.

A school was deemed to have polarised if the difference between the LEA mean index score and the 'actual' intake ('after choice') index score was *larger* than the difference between the LEA mean and the 'all catchment' intake ('before choice') index score. If these comparisons with the LEA mean were the same, or if the difference between the LEA mean and the 'actual' intake profile was the smallest of the two, then the school was defined as having not polarised.

Social Polarisation of All School Intakes

Using this binary calculation of social polarisation for each school, and for each of the 11 GB Profiler '91 variables, the Table 7.5 was produced. This shows the proportion of schools within each Local Authority, against each of the variables, that were defined as having polarised. Across all the study LEAs, and across all the variables, this table shows that 38.86% of the schools had polarised to some extent. Conversely, this meant that over 61% of schools had not polarised and, either, remained as different from the LEA mean, as they would without open enrolment, or they actually homogenised, i.e. became similar to the LEA average. So, for the majority of schools, the socio-economic composition of their intakes had become similar to the average LEA socio-economic profile. But, on the other hand a fairly significant minority of schools polarised, to some extent, away from the LEA average.

Table 7.5 also shows that there were some significant variations in the proportion of schools that polarised by both variables and LEAs. For example, the variables that highlighted the greatest polarisation were white-collar workers, younger people, privately owned and wealth indicators (all showed over 40% polarisation). All four of these represented a spectrum of socio-economic characteristics that could used be to indicate social 'advantage' or 'disadvantage'.

Across the Local Authorities the contrast between the Inner London borough (East) and the Eastern county LEA was quite considerable. Only 27% of schools in the London LEA polarised, but in the county LEA just over half of all schools polarised. The difference between these two Local Authorities could not be more pronounced. For the schools in the county LEA, every variable showed a degree of polarisation above the overall average, whereas, for the schools in the Inner London borough, only one variable (blue-collar workers) showed the extent of polarisation seen across the whole of the study regions.

Table 7.5   Proportion of schools that polarised by LEAs and by GB Profiler '91 variables

Variables	Proportion of schools (%)			
	Inner London Borough (East)	Large Metropolitan Borough (West Midlands)	Small Metropolitan Borough (West Midlands)	Metropolitan Borough (Greater Manchester)
Younger People	26.67	33.33	65.00	45.00
Older People	26.67	36.67	50.00	30.00
Ethnicity	13.33	26.67	30.00	25.00
Privately Owned	33.33	35.00	50.00	30.00
Rented	33.33	28.33	60.00	35.00
White-collar Workers	26.67	45.00	55.00	30.00
Blue-collar Workers	40.00	30.00	30.00	35.00
Young Families	26.67	31.67	45.00	30.00
Older Families	26.67	31.67	45.00	35.00
Wealth Indicators	26.67	31.67	50.00	30.00
Poverty Indicators	20.00	28.33	50.00	35.00
All variables	*27.27*	*32.58*	*48.18*	*32.73*

This was not just an urban-rural contrast, since the urban small metropolitan borough in the West Midlands also had a high degree of social polarisation overall (48.18%), and urban schools within the Eastern county

Table 7.5   (Continued)

Variables	Proportion of schools (%)		
	County (Eastern)	County (West Midlands)	All schools in study
Younger People	44.74	44.83	*41.76*
Older People	47.37	41.38	*39.56*
Ethnicity	57.89	37.93	*34.07*
Privately Owned	47.37	44.83	*40.11*
Rented	52.63	37.93	*39.56*
White-collar Workers	55.26	48.28	*45.60*
Blue-collar Workers	55.26	31.03	*36.81*
Young Families	42.11	27.59	*34.07*
Older Families	42.11	44.83	*37.36*
Wealth Indicators	57.89	41.38	*40.11*
Poverty Indicators	50.00	48.28	*38.46*
All variables	*50.24*	*40.75*	*38.86*

LEA also polarised. Instead, the number of schools that had polarised within each authority was related to their respective level of market activity and competition. Figure 7.4 illustrates the relationship between a measure of market activity and the proportion of schools that polarised. This suggested that as more students participated in the market then the allocation of students to schools was less likely to follow distinct social cleavages.

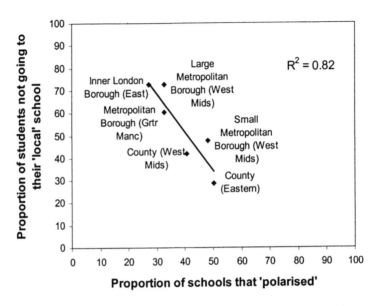

Figure 7.4 **Relationship   between   market   activity   and   social polarisation of school intakes**

The reason why a number of Local Authorities appeared to have less overall market activity and competition was discussed in Chapter 3, but, briefly, this was probably a combination of the geographical context and the decisions by the LEAs in administering admissions procedures. The relative significance of these two factors did not hide the obvious conclusion that, the more constraints there were within the market place, the greater the likelihood that schools became more socially polarised and segregated. However, this argument could be over-simplified since it did not take into account to what extent, or why, schools were becoming more socially polarised. The discussion now turns to the rates of social polarisation within the new education market.

Rates of Social Polarisation of School Intakes

Taking only the 'occurrences' where schools had polarised, the average rates of polarisation were calculated. The total number of possible 'occurrences' was equal to the number of schools multiplied by the 11 variables being used. Consequently, a single school could have been counted as having polarised several times if it had polarised by more than one of the variables. The overall rate of polarisation was equal to 12.18, which was, subsequently, used as a relative point from which to compare the rates of polarisation by LEA and by variable. However, the size of this average polarisation was worthy of comment. Since GB Profiler '91 used an index score, in which the UK average for each variable was equal to 100, an increase of 12.18 in either direction from the LEA mean did not seem great. This was particularly true when one realised that the standard deviation of all index scores for all variables and for all pupils was equal to 71.5. What remains to be a large, or significant, rate of social polarisation is debatable, but this average did give some measure to compare the individual rates of polarisation.

The variables that seemed to indicate the greatest rate of social polarisation were ethnicity, wealth indicators and blue-collar workers. The ethnicity variable, however, only showed high levels of polarisation in the large West Midlands metropolitan borough, the small West Midlands metropolitan borough, and, to some extent, the Inner London borough (East). This probably reflected the greater presence of multi-ethnic populations within these Local Authorities, but the large rates of relative polarisation only highlighted the great significance ethnicity has on the market place.

The other variable, which generally showed an above average rate of polarisation, was the blue-collar workers index. On closer examination it was clear that this variable was really only significant in the two more rural Local Authorities, in the Eastern and West Midlands regions of England. This was probably because of the very high index scores for this variable in farming communities. This led to the noticeable differences between the variables that highlighted polarisation in the urban and rural context. In the rural Local Authorities it was only the wealth Indicators and the blue-collar workers that appeared to show any great rate of social polarisation. This contrasted quite significantly with the more urban authorities, in which nearly all the variables for social poverty showed high rates of polarisation in one authority or more. This probably reflected the greater social diversity to be found in urban landscapes.

The rates of social polarisation differed quite markedly between each of the Local Authorities in this study. Both the small and large metropolitan boroughs of the West Midlands had high rates of polarisation, whereas the Inner London borough (East), and the two rural county LEAs had relatively small rates of polarisation. However, the rate of polarisation did not seem to be related to, either, the proportion of polarising schools, or the level of market activity. For example, the Inner London borough (East) had a small amount of schools polarising and a relatively small rate of polarisation. Conversely, the small West Midlands metropolitan borough had a large number of schools polarising, and it appeared that these were polarising at quite a significant rate. It was unclear from this as to why the rates of polarisation might have differed, but it did show that the amount and rate of polarisation needed to be considered together in order to really identify the degree of social inequality and social exclusion in the market place.

This has shown the variation in the relative amount of social polarisation occurring in the market place, and from this it is possible to identify the relative degree of social exclusion that exists in the market place. However, as outlined in earlier schools tend to compete in different ways. Therefore, what is the relationship between the different competitive positions schools have in the market place and the patterns of socio-economic polarisation of schools?

Competition Spaces of Polarised Schools

In Chapter 4 competition spaces were defined as areas of interaction between schools in the market place, and from this definition each school within the study was classified according to their competition characteristics. Within these competition spaces, schools competed with other schools in different ways. A typology of competition was thus developed, using the following categories of school competition:

i.   'Private' competition spaces – diversity in the market place
- A - Competition extends across more than half of the LEA territory
- B - Competition is significantly spread but less than half of the LEA territory
- C - Competition is relatively local in nature
  - Ca - Cannot compete with better[a] state schools
  - Cb - Can compete with better[a] state schools
- D - Grant Maintained schools competing in a 'state' hierarchy

ii.   Hierarchical competition spaces
   - Ea - Top of a hierarchy
   - Eb - Top of a small hierarchy [b]
   - Fa - Middle of a hierarchy
   - Fb - Lower middle of a hierarchy
   - Ga - Bottom of a hierarchy
   - Gb - Bottom of a small hierarchy [b]

iii. Non-hierarchical competition spaces
   - H - Equal competition between schools
   - J - No competition between schools of the same 'type' (i.e. parallel competition only)

a   Based on the GCSE examination performance of the schools.
b   A hierarchy made up of only two schools or with a relatively small amount of competition between schools.

One of the important outcomes of these competition spaces was to see if and how the socio-economic composition of intakes had changed depending upon which type of competition space, and position within it, that a school was located. Table 7.6 shows the proportion of polarising 'events' and the average rate of polarisation for each type of school competition.

It is clear from this table that there was the same likelihood of schools, between 30 and 39%, to polarise in each of the competition categories. The obvious exception to this was the very high proportion of polarised schools (52.83%) that were described as being at the bottom of 'state' hierarchical competition spaces. It was these schools that were clearly not successful in the education market.

The average rates of polarisation varied much more between the different categories of school competition (private, state, or non-hierarchical). The schools that had polarised at the greatest rate were the 'private' schools. However, there appeared to be a distinction between the different types of 'private' competition spaces. Therefore, the 'private' schools that attracted pupils from across more than half of the Local Authority polarised at a high average rate of 20.37. But those that attracted pupils from less than half of the Local Authority only polarised at an average rate of 10.66. Again, this latter type of 'private' competition space differed from 'private' schools that appeared to be competing with 'local' state schools. These, more locally competitive, 'private' schools polarised at a rate similar to the most attractive of 'private' schools.

Table 7.6   Competition space types and intake polarisation

Competition Space Type[a]	Proportion of 'events' that polarised (%)[b]	Average rate of polarisation
A (Private)	32.39	20.37
B (Private)	29.70	10.66
C (Private)	34.97	23.87
E (a or b) (State)	34.44	8.80
F (a or b) (State)	38.64	11.44
G (a or b) (State)	52.83	10.18
H or J (Non-hierarchical)	39.26	4.94

a   Type D of school competition (GM schools competing locally) has been omitted because there were only a few schools that had been categorised as such.

b   A polarising 'event 'occurs when a school appears to have moved away from the LEA average for each and every socio-economic variable considered. Therefore the maximum number of 'events' is equal to the number of schools multiplied by the 11 variables.

This distinction between 'private' schools was evident when comparing the average rate and direction of polarisation for each of the GB Profiler '91 variables. Figure 7.5 shows that the category A schools and category C schools polarised in the opposite direction to one another. Generally, the 'private' schools that attracted pupils from across more than half of the Local Authority polarised their intakes with children from socially 'advantaged' backgrounds, whereas, the intakes of the 'private' schools that competed locally, became more concentrated with children from more relatively socially 'disadvantaged' backgrounds. However, caution should be made about any conclusions that arose from this, since polarisation was calculated by comparing the catchment intake with the actual intake. In the context of the more 'private' schools there could be no assumption that, the intakes *before* open enrolment was introduced, actually reflected the

catchment intake at any time. It did suggest that, however, since many of the locally competing 'private' schools were located in relatively prosperous areas, their intakes did not exactly match the socio-economic characteristics of their locale.

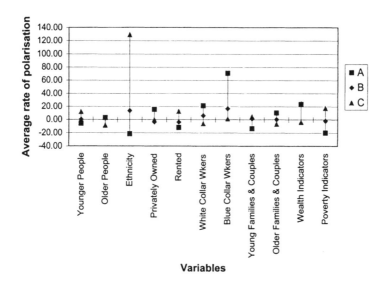

Figure 7.5   Comparison of rate and direction of polarisation of schools in 'private' competition spaces, A, B and C, by GB Profiler '91 variables

Within the 'state' hierarchical competition spaces (types E, F and G) there were greater similarities (Table 7.6 and Figure 7.6). However, it was interesting to note that, the schools that appeared to be the most 'attractive' in 'state' competition spaces, typically polarised at a smaller rate than any other category of schools. This contrasted quite significantly with schools from the bottom of 'state' hierarchies, which, along with having a greater propensity to polarise, did so, on average, at a relatively high rate.

Considering that both, Table 7.6, and the comparison between Figure 7.5 and Figure 7.6, showed that the rate of polarisation for 'state' schools, in general, was not as great as it was for 'private' schools, the difference between the polarisation of schools at the top and of schools at the bottom of 'state' hierarchies was quite significant. Figure 7.6 shows that, generally, the intakes of the more *attractive* schools were becoming increasingly concentrated with children from socially 'advantaged' backgrounds. The least *popular* schools, however, were increasingly comprised of children from socially 'disadvantaged' backgrounds. Schools that were found to be in the middle of 'state' competition hierarchies were affected in a similar way to schools at the bottom of 'state' competition hierarchies.

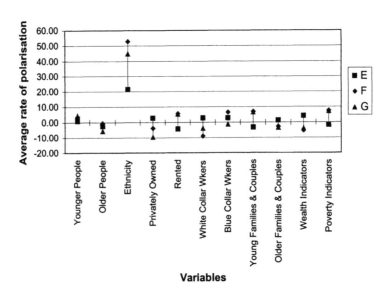

Figure 7.6 **Comparison of rate and direction of polarisation of schools in 'state' competition space, E, F and G, by GB Profiler '91 variables**

The analysis in Chapter 4 showed that the position of schools within local 'state' hierarchies was related to their examination performance relative to the performance of other schools within their competition space. Consequently, the contrast in the social polarisation between schools at the top, and schools at the middle and bottom, of the local 'state' hierarchies, was similar to the findings of polarisation between schools with the highest and lowest examination performance.

The final set of 'state' schools, those that were deemed not to be in a hierarchical competition space had, as perhaps expected, the smallest rate of polarisation of all the schools. However, even though these schools were, supposedly, not in any great competition with other schools, they still had a similar probability of polarising as schools in other types of competition spaces. Of the schools that did polarise, their intakes became generally more concentrated with children from socially 'disadvantaged' backgrounds. But, as suggested, the *rate* of this polarisation was not that significant.

This analysis of the social polarisation of intakes, according to different characteristics of schools, has shown that, again, the most significant changes in the composition of intakes were for particular *types* of schools, and did not develop an overall pattern of change. So, for example, there was little significant social polarisation of intakes for 'private' schools altogether, but when the 'private' schools were divided up into their different respective competition spaces, the conclusion was very different. The 'private' schools, which attracted pupils from across a significant area of the respective Local Authorities, had, on average, the greatest rate of social polarisation than any other type of school. These particular schools appeared to be attracting children from the most affluent areas in the Local Authorities, and this was being represented in the overall socio-economic composition of their intakes.

For schools from the 'state' end of the private-state continuum there were also shifts in the intake composition of particular schools. The majority of change was in the intakes of schools that appeared to be the least effective, or least 'popular', in the market place. These same schools generally had the worst examination performances, both across the Local Authority, and within their competition spaces, and, became increasingly concentrated with pupils from relatively socially disadvantaged backgrounds. This contrasted quite sharply with many of the schools from the top of local 'state' hierarchies, particularly those with the highest examination performance within their respective Local Authorities.

The different types of schools that were polarising also appeared to be polarising for different reasons. For example, the 'private' schools were obviously attracting a particular set of pupils, and were, therefore, becoming polarised because of this *new*, 'external', section of their intakes. On the other hand, 'state' schools with the lowest examination performance were not attracting students, and, therefore, must have been polarising because a particular set of their 'local' catchment pupils were going elsewhere. The following section addresses the different underlying causes of social polarisation.

Causes of Social Polarisation

The previous discussion began to highlight the difference between, what could be called, 'direct' and 'indirect' polarisation within the education market. *Direct polarisation* occurred when particular sets of pupils were attracted to a particular school. Consequently, the social polarisation of such intakes was determined by the intake from outside of the catchment area. *Indirect polarisation* differed from direct polarisation in that the social polarisation occurred because particular sets of pupils were being 'left behind' within a particular school. In this case the social polarisation was determined largely by the socio-economic characteristics of the 'local' catchment pupils who were being 'lost' to another school.

Using the five different intake profiles defined in Chapter 5, the effects of the 'outgoing' students ('lost' intake profile) and the 'incoming' students ('outside' intake profile) were compared with the students who continued to attend their 'local' school ('catchment' intake profile). Having studied the intake profiles for all schools that had significantly polarised (i.e. above the average rate of polarisation), nine different causes for polarisation were identified (Figure 7.7). These formed a continuum of importance between the 'outside' intakes (*direct polarisation*) and the 'lost' intakes (*indirect polarisation*). Across these nine different causes of social polarisation it was recognised that there were four different levels at which the outside and lost intake played a part in the social polarisation of the intakes:

i.   Primary cause of polarisation.
ii.   Secondary cause of polarisation.
iii. Perpetuating polarisation.
iv.   Correcting potential polarisation (three levels of this).

As Figure 7.7 shows, the causes of social polarisation were skewed, in that *direct polarisation* was more predominant than *indirect polarisation*. The potential for polarisation was greater in the middle of the continuum, since this was where both causes of polarisation had a primary role in altering the intake composition. At both ends of the continuum, where only one cause was the primary cause of polarisation, the other cause began to compensate for the change in the intake. For example, if affluent families were sending their children to a particular school (*direct* = primary), **and** less affluent 'local' families were going to an alternative school (*indirect* = primary), then these conditions would have enormous potential to create social polarisation. If, however, it were not the less affluent 'local' families sending their children to an alternative school, but instead *affluent* 'local' families were the ones leaving, then the 'new' intake from outside would simply be replacing those that were leaving. In this case the *indirect* cause of polarisation would be compensating for the changes to the intake and therefore reducing the potential for social polarisation.

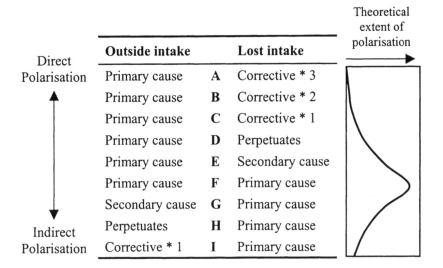

Figure 7.7   Causes of social polarisation in school intakes

The proportion of actual polarisation (Figure 7.8) did relate to the potential for polarisation as indicated in Figure 7.7. However, the most significant cause of school polarisation occurred when the primary cause of polarisation was *direct* and when the *indirect* cause was of secondary importance.

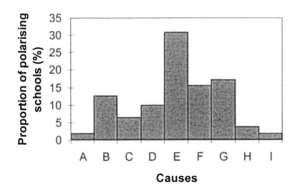

Figure 7.8   Proportion of polarising schools by category of cause* of polarisation

* See Figure 7.7 for definition of causes.

The cause of polarisation was also directly related to the actual changes to the socio-economic characteristics of the intakes for schools that polarised. Figure 7.9 shows the distribution of schools for each cause of polarisation depending upon whether their intakes were becoming more socially 'advantaged' or more socially 'disadvantaged'. Changes to the socio-economic composition of intakes due to *direct* polarisation alone, generally became more socially 'advantaged'. Conversely, changes to the socio-economic composition of intakes due to *indirect* polarisation generally became more socially 'disadvantaged'. This supported the argument that schools becoming more concentrated with children from socially 'disadvantaged' backgrounds were losing the relatively socially 'advantaged' pupils.

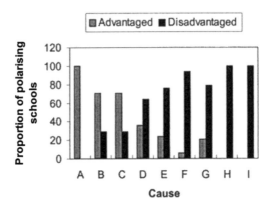

Figure 7.9 Proportion of polarising schools according to whether they were becoming more socially 'advantaged' or socially 'disadvantaged' by category of cause* of polarisation

* See Figure 7.7 for definition of causes.

A key factor to emanate from this analysis was the way in which the direct causes of polarisation and indirect causes of polarisation were occasionally compensating for each other and, consequently, reversing, or correcting, the polarisation of such intakes. For example, Figure 7.10 illustrates the importance of each cause of polarisation for schools in 'private' competition spaces. It had already been ascertained that schools with type A competition, where they attracted pupils from across more than half of the LEA, had very high rates of polarisation. But, Figure 7.10 shows that, in the majority of cases, the effects of **not** attracting some of the 'local' pupils was actually tempering the rate of polarisation. If these schools took more of their 'local' pupils, along with pupils from across the Local Authority, then the polarisation would, potentially, have been even greater.

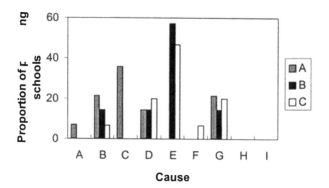

Figure 7.10 Proportion of polarising in 'private' competition spaces, A, B and C, by category of cause* of polarisation

* See Figure 7.7 for definition of causes.

This might seem a rather unnecessary conclusion to have made, since 'private' schools, and particularly those with type A competition, would have had similar intakes before Open Enrolment was introduced. However, similar observations were made for schools in local 'state' hierarchical *competition spaces.* Figure 7.11 shows the importance of each cause of polarisation for schools at the top of the local 'state' hierarchies. The corrective processes of some 'local' pupils going to an alternative school were also limiting the changes to the intakes of a significant proportion of these schools. This would suggest that the results prevented in this Chapter with regards to the extent of social polarisation of intakes might were hiding the potential for greater social polarisation.

Identifying the causes of social polarisation in school intakes highlighted that intakes of 'unpopular' schools were indeed polarising because a particular subset of 'local' pupils were being 'left behind' in these unpopular schools. At the other end of the market performance scale there were 'popular' schools that were socially polarising because they were attracting a particular subset of pupils from outside the 'local' area. In both cases these subsets of pupils that were identified, followed traditional social class cleavages. Pupils from 'disadvantaged' backgrounds were being left behind in poor performing schools, and pupils from 'advantaged' backgrounds were getting into schools that already had pupils from relatively privileged backgrounds.

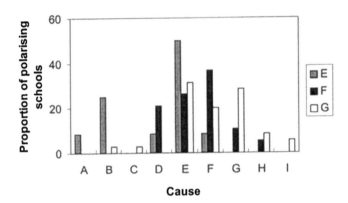

Figure 7.11 Proportion of polarising in 'state' hierarchical competition spaces, E, F and G, by category of cause* of polarisation

* See Figure 7.7 for definition of causes.

These observations, however, need to be put back into context against the other findings from this Chapter, particularly the fact that schools, which had socially polarised, were in the minority of all schools. In order to address all the issues presented here the following section discusses the findings within the context of the social inequalities that may have arisen.

Conclusions

This analysis on the outcomes of school choice in education has uncovered new conclusions, and found empirical support for others. It has also provided a useful platform from which to examine the effects of the 'new' education market upon social mobility. However, within the scope of this research it was only possible to examine the processes of exclusion, and the resulting social inequalities that arose within the market place, in order to posit ways in which they might have led to more general consequences for social exclusion and social mobility.

This study has produced evidence to support the argument that parents and children from all social classes participate in the market place to some extent. The complexity of school choice also proved that the only comparison to be made between different levels of engagement was in the **probability** of particular social classes being 'active' or 'inactive'. In other words, there were no absolutes. The context to this was also very important, particularly the general levels of market activity and competition within each Local Authority. So, for example, there were parents from relatively 'disadvantaged' backgrounds who were more 'active' in the market place than parents from relatively 'advantaged' backgrounds. The key difference to these two scenarios was that the socially 'disadvantaged' parents lived in LEAs where there was greater competition between schools overall. The socially 'advantaged' parents, conversely, lived in LEAs with little competition and little encouragement by the Local Authority to be 'active' in the market place.

The findings have also shown that there was a small minority of pupils who were either, *very* 'active' in the market place, or non-participatory when everyone around them was 'active'. It was the actions of these parents who arguably powered social polarisation of intakes for a few schools. And only 'a few' schools it was, for it has been shown that the majority of school intakes were not becoming socially polarised. At the LEA level it was seen that the variation in intakes across Local Authorities was, overall, becoming more homogenous. Taking schools individually, the findings suggested that a significant minority of school intakes, just over one in every three schools, were polarising to some extent away from the LEA average socio-economic characteristics.

On close examination these schools tended to represent the extreme cases of competition typology; the most 'popular', and the least 'popular', schools. This analysis has shown that they were also represented by their position within the examination league table for their respective LEAs. However, this might have been expected since the typology of schools within competition spaces, as outlined in Chapter 4, tended to reflect their respective examination performances.

Even though it has been argued that for the majority of schools their intakes had actually homogenised towards the average socio-economic characteristics of their respective LEAs, concern must be raised that schools that had polarised represented the extremes in both measures of school performance. This was particularly so because the intakes of the top performing schools appeared to be increasingly concentrated with children from relatively more privileged backgrounds, whereas, the intakes of the lowest performing schools were becoming increasingly concentrated with children from relatively less prosperous and less 'advantaged' backgrounds.

It was also clear that changes to the intakes for these two sets of schools were due to slightly different reasons. Social polarisation of intakes for the lowest performing schools was largely due to the non-participation, or inability, of, a small number of individuals to send their children to an alternative school, and a reluctance of parents with relatively higher social 'advantage' to send their children to these schools. This contrasted quite sharply with the underlying cause of social polarisation for the higher performing schools. Here, social polarisation was being led by a few, but very 'active', parents, who came loaded with relatively high levels of social and cultural capital, perhaps necessary to get their children in to these more popular schools.

These results suggested that there were three tiers of outcomes for schools in the 'new' education market (Table 7.7), each with different levels of *performance*, both in terms of market and pedagogical performance, different changes in the *socio-economic characteristics* of their intakes, and different underlying *causes* of these changes.

The top tier and bottom tier of outcomes are inextricably linked to social inequalities, most evident in the level of resources schools would get from changes to the school rolls, and the effectiveness of the schools in gaining academic credentials for their pupils. These inequalities should also be added to the effects of polarisation on the socio-economic characteristics of intakes in reducing social integration and creating an environment that reproduces social and cultural norms and preferences.

However, two further points have to be raised that question these conclusions. The first is the need to take into account the spatial constraints upon the market place. These have been shown to hinder the overall development of market activity and competition, particularly across rural Local Authorities. The spatial arena also controls and determines the competition spaces at which the majority of changes to intakes take place. Therefore, it must be remembered that the association between market performance and examination performance is only made within the relative position of schools within **each** competition space. Even though similar patterns could be observed when comparing the highest and the lowest performing schools in each Local Authority examination league table, the direct exchange of students between these schools was not guaranteed. It was as though the market place acted as a large filter, in which, at every level, there was some social sorting within schools, but which did not really manifest itself until the filtering process reached the two ends.

Table 7.7   Three tiers of outcomes in the 'new' education market

Tier	Performance	Social class of intakes	Cause of change in intakes
Top	Very attractive and high examination performance*	Increasingly prosperous backgrounds	Led by very 'active' parents who have high levels of social and cultural capital
Middle	Fairly attractive and average examination performance*	Intakes becoming more alike (homogenising)	Varying levels of parental activity in the market place that seems to cross social class cleavages
Bottom	Not attractive and low examination performance*	Increasingly disadvantaged backgrounds	A combination of some parents, with relatively greater social and cultural capital, sending their children elsewhere, and parents who are not participating in the market place, who tend to have relatively low levels of social and cultural capital

*   Relative to other schools in their respective 'competition space'.

This, however, leads to the second point and asks how many schools, at the two ends of the filter, bear these effects, since, there is, in theory, only one school at each end of the performance scales. The findings in this Chapter suggested that approximately 38% of schools from either end polarised, to some extent. In reality, schools that had significantly polarised were fewer in number than this 38% suggested, and tended to be from the bottom tier rather than the top tier.

The market has clearly worked against these schools and the pupils who continue to attend these schools. But, it was LEAs with the *least* market development that tended to have the greatest social polarisation of intakes.

The market also managed to homogenise the majority of school intakes to such an extent so that, across the whole of the Local Authorities, there was less variation between the overall socio-economic school profiles. And, once again, the greatest homogenisation was in the Local Authorities with the greatest level of participation in the market by parents and schools.

Evidently, there were two sides to the social equity of the 'new' education market. On the one hand it could be argued that the market makes access to schools more equitable. This can be seen in this study, and others, where the socio-economic composition of intakes is becoming more integrated for many schools, more integrated and mixed than they would have been if catchment areas were being used. The question is, are these schools part of some greater market process in which there will always be 'winners' and 'losers' in the market place. The research presented here suggested that there were some 'winners' and 'losers', both for schools and for parents. And these were indeed polarising away from each other in terms of academic performance and by social class. This was the inequitable side of the education market.

If these two scenarios were to be accepted, then it could be argued that the 'new' education market is unjust. However, this study provides two stokers to further fuel the debate. One is that the market has been equitable for the clear majority of parents at the expense of the few. Within the debate on social justice this might be seen as just. The second is that it has shown that the market appears to be inequitable in varying degrees according to the development of different market places. This simply reminds us that this is a dynamic and changing market and that the equity of the market will also continue to change. The final Chapter in this study will attempt to address these two issues further.

Notes

1 This suggests that attending a school that already has a high proportion of its pupils obtaining good examination performances will improve the chances of a child in achieving the most of their ability in their examinations. See Willms and Echols (1992) for a support for this proposition.

2 See Taylor (2000) for a greater discussion of these methods.

8 Conclusions

Introduction

The main aim of this study has been to provide a geographical interpretation of the 'new' education market in England and Wales. It attempted this by undertaking a comparative study of the patterns and processes of competition and choice across different geographical contexts – institutional, spatial and social. As discussed in the previous Chapter, in relation to all other studies on the 'new' education market, this research focussed on issues of social justice, both on the process-side and the product-side of the market mechanisms. This study has raised many issues, both in developing an understanding of the geography of the 'new' education market and in identifying areas for future research. This final Chapter summarises the key geographical and social variations of the education market identified in the research before evaluating the *quasi-market* approach to education provision in England and Wales. Since this study focussed on issues of social justice rather than economic efficiency of the reforms, the conclusions also focus on issues of equity and equality. The Chapter then concludes by identifying some of the most important areas of the 'new' education market still in need for further research and which pose potential policy dilemmas.

The Geography of the 'New' Education Market

An implicit aim of this research has been to develop the notion of the geography of the 'new' education market. Other research has often discussed the socio-spatial dialectic as a tool of enquiry (Gewirtz *et al.*, 1994), the geography of the 'new' education market develops this relationship between society and space. This presentation of the geography of the 'new' education market focuses on two key sets of variations: spatial and social.

Spatial Variations

There have been two key spatial features that have arisen from this research. The first was the geographical variation of the education market at the LEA, or market place, scale. The second was the significance of space upon the operation of market mechanisms and competition between schools. These two features are now summarised in turn.

This research has shown the enormous variation in the levels of market activity that took place in different LEAs. These variations were seen in the proportion of pupils who attended a school other than their nearest, or 'local', school, the differing levels of engagement parents had with the market place, and the extent to which parents overcame proximity and accessibility to send their children to particular schools. Consequently, some LEAs provided examples of considerable market activity in the form of choice by parents, and competition between schools, while others only showed only limited deviation in the process of school choice and competition that existed before the reforms of the last twelve years.

There were two important sets of factors that seemed to determine these variations. The first were geographical and related to the spatial distribution of schools and pupils alongside the socio-economic and cultural characteristics of parents in the process of choosing a school. The second set of factors were institutional and tended to relate to the administrative and organisational duties of Local Authorities, both at the time, and as a culmination of historical decisions made over the last 60 years. In particular, the admissions policies of LEAs at the time, and in the recent past, played a role in determining the levels of parental engagement with the market, along with the development of schooling and the diversity offered to parents as a result of changing legislation and building programmes.

These variations in levels of market activity were significant, since they were related to differing rates of participation of parents from different socio-economic backgrounds. So, for example, it was the most socially 'advantaged' parents who were the most likely to use their powers of school choice at the *earlier* stages of market development within a LEA. Local Authorities that were able to offer and encourage greater choice were, consequently, the ones with greater participation, and therefore, more inclusive, of parents from different socio-economic backgrounds.

However, it was also shown that the number of appeals lodged across different LEAs was significantly related to the levels of market activity. Hence, with greater participation came greater frustration and dissatisfaction with the resulting allocation of pupils to schools. This leads to the great importance that the number of surplus places had in the education market place. With some surplus in the system there is the opportunity to sift and sort pupils into different schools. But, without that surplus there are, first, greater administrative problems for the LEA admissions teams, and second, some schools become oversubscribed and unable to meet the demand for places.

The variation in the number and proportion of appeals lodged in each LEA was also related to the private-state continuum, or diversity of schools, of the authorities. It was shown that the importance of the 'private' sphere of education, even within the LEA-maintained sector of schooling, contrasted between LEAs. This generated different profiles of the private-state continuum for each authority. There were two interpretations of the impact of the 'private' sphere upon the number of appeals lodged. On the one hand, the more 'private' the schools were in an LEA the greater the levels of *exclusion* within the choice process, and, as a result, greater frustration and dissatisfaction with the process. On the other hand, it was suggested that the presence of the 'private' sphere within the market place was creating greater market activity by parents and schools. This was because it established greater market development and competition between schools, increased parental empowerment, and encouraged greater 'private' consumption, and behaviour, of parents for traditional 'state' goods. These two sides to the presence of the 'private' sphere in schooling were both seen in parents' responses to diversity in the market place. Some parents saw diversity as providing greater choice, but others thought that it actually limited their choice of schools. This contrasting view is largely a product of diversity in the English and Welsh education system being closely linked to the 'private' sphere of provision.

These spatial variations between LEAs, or market places, were significant because they determined two key components of the education market: the participation and competition in the market place, and the diversity of choice that both constrained and provided choice for parents. LEAs provide different contexts from which the market place operates, and are, therefore, influential upon the consequences of the 'new' education market.

The spatial relationship between competition and choice, or the 'lived' market place, was also necessary to appreciate before interpreting the consequences of the market place. It was very rare for all schools in a LEA to be directly competing against one another, just as much as it was difficult to find parents who chose from across the whole authority. Subsequently, *competition spaces* were an important component of the 'lived' market place. Competition spaces proved to be very complex since they were often connected *in some way* with other competition spaces and did not form discrete areas or 'local' markets. They were relatively small, but they varied in size, generally encompassing between four and nine schools. This variation was reflected in the number of schools that parents themselves considered when choosing a school, between one school and seven schools. The size of such competition spaces was also related to the spatial nature of the authority; they were generally comprised of more schools in urban areas than they were in rural areas. Competition spaces in urban Authorities were also more complex and difficult to disentangle.

This research identified three main forms of competition space; 'private', 'state' hierarchical, and non-hierarchical. 'Private' competition spaces appeared to operate in 'parallel' to one another and to the other two forms of competition space. However, they did impact, to varying degrees, on these other 'parallel' markets. This was also seen in the *types* of schools that parents considered when choosing a school. There was often some overlap in the types of schools that parents chose from, but there were also clear distinctions, suggesting that parents did engage with different 'parallel' markets.

The most frequently occurring form of competition was in 'state' hierarchical competition spaces, which positioned schools, typically, by their examination performances relative to other schools in the same competition space. These were more localised than 'private' competition spaces and usually contained three or four schools, even though there were many examples where several 'state' hierarchical competition spaces were connected. Again the schools considered by parents during the decision-making process reflected these positions since they never considered schools that were below the relative position of their 'local' catchment school. Consequently, none of the parents in this study chose a school at the bottom of a local 'state' hierarchy unless it was, in fact, their nearest school. This research also highlighted that, given the parents concern for *process* criteria, it was school reputations, determined locally, and school examination performances, that provided the crude basis for their decision-making. Even though parents generally tried to play down the role of these two *product* criteria it was impossible to ignore the significant relationship between the structure of the hierarchical competition and the schools relative examination performance.

The non-hierarchical competition spaces were often not actually spaces, for they included 'state' schools that were in no 'direct' competition with other 'state' schools. These schools were generally geographically isolated in the market place. The important feature of this isolation was that it occurred in both urban and rural locations. Such market isolation was relative to the geography of the home area. Hence, to be isolated in a rural area probably meant that a school was, say, 10km from another school, while in an urban area an isolated school with no competition was, probably, only 2km away from other schools. Parents highlighted this relative market isolation, since it was those, attending schools in non-hierarchical competition spaces, who stressed the greatest constraints upon their decision-making.

Cutting across these different competition spaces were a number of geographical characteristics such as the schools' relative location to other schools in their competition space. There were many cases where the uneven playing field of competition spaces was impacting upon the levels of competition and choice for particular schools. For example, schools that were centrally located in their competition space, or which were more accessible than others, such as schools in small rural towns, often appeared to 'perform' better in the market than their relative position within the examination league table might have suggested. Conversely, schools that were located in peripheral areas of their competition spaces found it difficult to appear attractive, and, therefore, able to increase the size of their intakes.

Another geographical characteristic of these competition spaces was the significance of the 'local' intake. Of all the schools in this study 36% of them had what was called 'prominent local intakes'. Such schools were those that appeared to be unpopular in the market place or those that were clearly oversubscribed and, hence, were forced to take pupils according to their proximity to the school. This had an important impact upon school-community relationships, for it was those parents who attended schools with intakes from a wide area that were the least inclined to identify their school as part of the community they lived in. Similarly, parents rarely identified schools that tended to be the least popular in the market place as being *local* to them, even if such schools were clearly their nearest school. This relationship between schools and their locale is very significant, if not just to the community, but also to how parents go about choosing a school for their children. This research often showed how the 'local' or nearest school was frequently used to contextualise their choice, either as a way of encouraging greater participation in the market place, or as a yard-stick from which to evaluate other schools.

Ultimately it was the spatial constraints on parents when they were choosing a school that were the most important in terms of equity in the market place. It was shown that the means of access to particular schools was closely related to the socio-economic characteristics of the parents. Pupils from relatively 'disadvantaged' backgrounds did have the means to travel to schools beyond their nearest school, but it was pupils from more 'advantaged' backgrounds that had access to particular means of transport that enabled them to cover greater distances, and, therefore, had more schools to choose from. There were examples of this in both urban and rural areas, and, even though different means of transport were used in these two types of areas, the relationship with social class and material wealth was equally as strong.

Overall it was shown that 40% of parents in this study thought that other parents did not have the opportunity or ability to choose in the 'new' education market. And in nearly all of these cases parents stated geographical and material reasons for these constraints. Within the geography of the 'new' education market this relationship between spatial inequalities and social inequalities is very important, so now the focus shifts to social variations in the market place.

Social Variations

This research has shown that overall 'between-school' social segregation across LEAs, the differences between the aggregated social status of school intakes, was actually less than it would have been if pupils were allocated to their nearest school. This illustrates the point that parents, from both middle-class and working-class neighbourhoods, were just as likely as each other to avoid their nearest school and choose an alternative school. The effect of this was for the aggregated social status of intakes to become more alike. However, the research suggested that there were some social variations in the *process-side* of choosing a school, and, therefore, variations in the degree of activity or engagement with the education market. The most 'active' in the market place were clearly the more socially 'advantaged' parents and, hence, their choice of schools was different. The consequence of this, on the *product-side* of the market, was for *some* social polarisation of *some* schools. That is, the social status of a few school intakes in each LEA actually became more distinct, with either higher or lower social classes, from other school intakes.

On the process-side of social inequality in the market place there were no statistically significant differences between working-class and middle-class families. However, there were varying tendencies of these two sets of parents to go through the decision-making process differently. So, for example, the role and importance of different members of a household tended to vary according to their socio-economic environment. It was seen how a child's input varied depending upon the social status of the family. But it also showed that there were two middle-class scenarios, one where the child's role was less important than in other households and one where the child was much more important. However, it was working-class families that tended to give the views of the child greater importance on the final decision. Overriding these social divisions of the balance of power within the household was the prominence of mothers in the decision of choosing a school.

Of greater significance to social cleavages, within the process-side of the market place, was the greater investment, both in time and energy, of some parents over others. As already indicated, parents from various backgrounds engaged with the market, but there was a clear distinction between the *most* and the *least* 'active' in the market place. This was reflected in the process of acquiring information on their chosen schools, in particular the difference between 'hot', or grapevine, information and 'cold', more official, information. The official sources of information, such as examination league tables and school prospectuses, were much easier to get access to but the least useful in aiding a decision. The grapevine was also used by nearly everyone, but the levels of engagement with the grapevines and the quality of information from such engagement varied according to the social status of parents. However, the most 'active' in the market place were less dependent upon social networks for their information since they arranged to visit all the schools they were choosing from and tended to access every piece of information equally. The ability of such parents was even more magnified since they also tended to choose from the most number of schools in the first place.

As a result of these small, but significant, social variations in the decision-making process there were three tiers of outcomes on the *product-side* of the market process. The majority of schools actually appeared to have more socially mixed intakes, but a few schools, which tended to be the best and worst academically performing, did have socially polarised intakes. These outcomes certainly reinforced the 'state' hierarchical competition spaces discussed above.

However, the caveat to these conclusions was that there was less social polarisation of school intakes in Local Authorities with greater overall activity by parents in the market place. In such areas the process of choosing a school was clearly more *inclusive* than it was in authorities with less opportunity, and with parents with less ability, to choose a school other than the nearest, or allocated, catchment school. This did not mean that the market was equitable in such LEAs, rather it meant that the market in these authorities was less inequitable.

Once again this began to show how spatial and social variations in the market place that comprise the geography of the 'new' education market are related to one another. Both the spatial and social variations in the market place combine to produce varying constraints and inequalities in choosing a school. For example, the spatial constraints can compound the social constraints and vice versa. However, the best example of how these two features of the geography of the 'new' education market combined was the greater participation of all parents in more developed market places where there was greater access to schools, and which, as a result, reduced the likelihood for social segregation between schools and social polarisation of school intakes.

These conclusions lead on to the issue of whether the market reforms to education provision in England and Wales has been 'just'. The market mechanism does, theoretically, allow parents to overcome spatial and social injustice, and with greater educational opportunities these reforms could enable upward social mobility. This research has shown that in reality this, to some extent, occurred. There were many parents from poor areas who sent their child to an alternative school to their local catchment school. Also, nearly every parent in this study believed that they should be able to make some preference over the school for their child, making it very difficult to remove this privilege from parents in the future. However, they were aware of its limitations and they stressed the great anxiety that was caused over these rights. But, whether these findings mean that the reforms are totally 'just' is debatable, since there are a few, but very significant, examples of schools that were socially polarising and which would have huge implications for their future academic performances. Using Rawls' (1972) theory of justice (also see Smith, 1994), based on a formulation of an egalitarian case where inequality is 'just' if it benefits all, especially the worst off, then it would be possible for the reforms to be 'just' if they did not make the position of the poorest pupils any worse off than they were without the reforms. Consequently, the fact that a few schools attract pupils who have the most privileged backgrounds are in turn the same schools that produce the highest academic credentials would be acceptable if the 'new' education market also benefited everyone else. This research has shown that in the majority of cases this was true. However, there were some pupils who, typically, continued to attend their local catchment school, which in turn had the lowest levels of examination performance in an area, and where other parents around them, usually those with relatively greater cultural, social and economic wealth, were sending their children to an alternative school. As a result, these schools were losing numbers and, subsequently, the resources to overcome the socio-economic problems that these children, in particular, had. Therefore, on the basis of this research the current reforms cannot be said to be 'just', not until the situation for these children is improved beyond the position they were in before the reforms.

Even if these reforms were adjusted to create a 'just' provision of education it would be wrong to assume that the system would make the provision of education any fairer. The best example of this is the rising conflict, particularly in urban areas, and which the media tends to focus on, between parents in the search for the best school for their children. The Greenwich and the Rotherham Judgements were the first of what could be many legal cases in which parents are taking Local Authorities to court over the allocation of schools. As some schools are becoming more 'attractive' the desire for parents to have the opportunity to send their children to these schools becomes more desperate. There is also the danger that the commodification of education will begin to reinforce, and possibly increase, residential segregation once admission policies for oversubscribed schools return to some form of catchment or distance rules (see the School Standards and Framework Act, 1998).

It is to the future of these education reforms that the conclusion now focuses on by identifying some of the key research and policy areas that, it is argued, need to be addressed.

Research and Policy Directions

The 'new' education market is almost certainly still under-researched. This study has begun to show the different complexities and dynamics of different Local Authorities and has provided conclusions that can be interpreted across many LEAs. Beyond deeper evaluation of the diversity of the education market future research must also draw in studies of school effectiveness, for it is the value a school gives to a child that determines social division and social mobility as the child progresses through adulthood. Another key element to future research should be to look at the consequences of the education market over time. This research, itself just a snapshot, has begun to show the differing effects of market places with varying levels of development, whilst a temporal approach would aid an understanding of the market processes and the way the market place is, itself, constantly changing.

These are some of the general routes that future research might follow in order to develop a greater understanding of, not just of the educational reforms, but also the use of quasi-markets in welfare provision. In terms of future policy direction, and the associated research that would inform that process, there are two routes it could take. In both cases the current set of reforms would be seen as being transitional. This is because there are three fundamental problems with the current market reforms that cannot continue in their present form. The first is that the majority of competition between schools, and the area from which parents are choosing schools, is too small to offer real choice and to encourage differentiation between schools. Bartlett (1993) argued that schools were no longer in monopolistic situations. However, particularly for relative isolated schools in both urban and rural areas, the schools have only moved as far as oligopoly environments, and might, therefore, not be fully operating under the pressures of a more developed market situation.

This is compounded by the second problem that, for the majority of 'state' schools, and possibly all the 'private' schools, the market is creating a hierarchical situation in which the schools are becoming either net gainers or net losers of pupils, where inflows are not necessarily being met with outflows, and vice versa. If these hierarchies are compounded year on year then this situation will reduce real choice and exacerbate conflict.

The final problem of the current reforms is the over-dependency of a system that is only offering diversity through 'private' and, therefore, exclusionary, means. This is, on the one hand, offering choice to some parents, but reducing choice for others. Real school autonomy should be encouraged within the 'public' sphere and not allowed to cross into the 'private' sphere.

On this basis it does not seem possible that the current reforms can continue without becoming more 'unjust' and creating greater conflict, because there will be little choice in a system that is based on consumer choice. In some respects this confirms what Hudson (1992) concluded about the quasi-market reforms in British health and social care. Hudson argued that the concept of the quasi-market was 'dubious' and 'ill-developed' (1992, p.140), and that greater regulation was needed. But, as Whitty (1997, p.33) stated, 'pointing to the damaging consequences of particular policies is not necessarily to question the motives of those proposing the reform agenda', and that a return to a bureaucratic system, which originally led to the reforms, could be an unattractive proposition.

The findings of this study, therefore, point to two possible directions in which the quasi-market of education provision can go in the future. In both cases the market structure would be seriously altered. The first route would be to allow the current changes to education provision in each Local Authority to develop to a situation where it becomes possible to identify schools for closure. The current market process has reduced the intakes of the most unpopular schools in each area and has therefore polarised the surplus in the education system. However, it must be acknowledged that unpopular schools are, generally, only relative to a *small* area, and are, therefore, not necessarily the most unpopular schools across the whole Local Authority. To close schools would require careful analysis of each school's performance over time, both in their examination performance and their market performance. Once the necessary surplus in each authority has been removed then the provision of places would have to return to some form of allocative system by the LEA. This does not necessarily mean that other areas of the reforms, such as school autonomy and devolved budgets, should also be removed. For example, a network of schools with specialisations offered to any pupils in any school within the network would maintain some diversity (Edwards and Whitty, 1997). However, such elements of the current reforms would be of less importance since they would not be in competition with other schools. The result of this shift would be a potentially 'new' look to the previous bureaucratic system in which administrators would be more responsive to the needs of pupils and schools, while still being able to plan efficient provision. Each authority would also have removed the least desired schools, which, in turn, would alleviate the pressure of parental dissatisfaction across the whole LEA. The danger of this route is that it could very easily fall back into the 'traditional' pattern of provision and encourage further residential segregation, particularly now that education has been commodified and parents have developed their consumer skills. Also, closing schools would not be an easy measure as there would be many social and cultural costs to the community that have been constantly overlooked by researchers and legislators (Dennison, 1983).

The second route for the reforms to proceed would be to keep Open Enrolment, but to encourage further market reforms. In particular, this would require a great deal of investment for extending access of choice to parents, such as providing free transport across areas larger than current competition spaces, and for enabling popular schools to increase their capacity through capital investment. Since the patterns of choice are currently hierarchical there would also need to be active encouragement of diversity in the market place that stays completely within the 'public' sphere. If parents have access to more choice, both numerically and by education type, then the presence of hierarchical choice could be reduced which would prevent social division and conflict. This would also require the relaxation of the National Curriculum to allow schools to develop their own personalities upon the market place. Another reason as to why the investment into school transport would have to be a necessary part of these reforms is that underlying demographic fluctuations in the pupil population (Thomas and Robson, 1984) would be exacerbated if provision was organised across small areas. In some respects these reforms would continue the quasi-market theme, but would also require greater regulation over the type of provision being offered within, and between adjacent, LEAs.

Both of these routes pose enormous advantages and disadvantages, but there does not appear to be any viable alternatives that would not be as fair or as 'just', and that would address the criticisms of the previous system of provision. The reforms have, thus far, brought about a change in provision not seen before, and have, therefore, created problems and dilemmas that need to be addressed, potentially, for the first time. The opportunity for continued change is there, but as highlighted by Whitty (1997) and Lauder (1991), legislators must develop a system of education based on democratic citizenship, where the balance between individual and collective interests and responsibilities is struck for greater equitable provision.

Conclusion

This research has hopefully extended the understanding of the education reforms over the last 20 years. It has also shown the importance of both a geographical and a more empirical study of such changes in provision. Overall, it has perhaps highlighted that the concept of a quasi-market in providing welfare systems is still in transition. Whether such a hybrid solution, or 'third way' can be maintained, or whether the neo-liberal or traditional welfarians will succeed in their lobbying, remains to be seen. The clear objective of the academic community is to continue to provide critical analysis of such legislation and to actively encourage education provision that is socially 'just' and of a high quality.

Bibliography

Abercrombie, N. and Warde, A. (1994), *Contemporary British Society* (2nd Edition), Polity Press, Cambridge.

Adler, M. (1997), 'Looking backwards to the future: parental choice and education policy', *British Educational Research Journal*, vol. 23, 3, pp.297-313.

Adler, M., Petch, A. and Tweedie, J. (1989), *Parental Choice and Educational Policy*, Edinburgh University Press, Edinburgh.

Ambler, J.S. (1994), 'Who benefits from educational choice? Some evidence from Europe', *Journal of Policy Analysis and Management*, vol. 13, 3, pp.454-476.

Audit Commission (1996), *Trading Places: The Supply And Allocation Of School Places*, The Audit Commission, London.

Ball, S.J. (1986), *Education*, Longman, London.

Ball, S.J. (1993), 'Education markets, choice and social class: the market as a class strategy in the UK and the USA', *British Journal of Sociology of Education*, vol. 14, 1, pp.3-19.

Ball, S.J. (1997), 'Policy sociology and critical social research: A personal view of recent education policy and policy research', *British Educational Research Journal*, vol. 23, 3, pp.257-274.

Ball, S.J. (1998), 'Big policies/Small world: An introduction to international perspectives in education policy', *Comparative Education*, vol. 34, 2, pp.119-130.

Ball, S.J., Bowe, R. and Gewirtz, S. (1994), 'Schools in the market place: An analysis of local market relations', in Bartlett, W., Propper, C., Wilson, D, and Le Grand, J. (eds), *Quasi-markets in the Welfare State: The Emerging Findings*, SAUS Publications, Bristol, pp.78-94.

Ball, S.J., Bowe, R. and Gewirtz, S. (1995), 'Circuits of schooling: A sociological explanation of parental choice in social class contexts', *Sociological Review*, vol. 43, 1, pp.52-78.

Ball, S.J., Bowe, R. and Gewirtz, S. (1996), 'School choice, social class and distinction: The realisation of social advantage in education', *Journal of Education Policy*, vol. 11, 1, pp.89-112.

Ball, S.J. and Vincent, C. (1998), ''I heard it on the grapevine': 'Hot' knowledge and school choice', *British Journal of Sociology of Education*, vol. 19, 3, pp.377-400.

Barry, N. (1991), 'Understanding the market', in Loney, M., Bocock, R., Clarke, J., Cochrane, A., Graham, P. and Wilson, M. (eds), *The State or the Market*, Sage Publications, London, pp.231-241.

Bartlett, W. (1993), 'Quasi-markets and educational reforms', in Le Grand J. and Bartlett, W. (eds), *Quasi-markets and Social Policy*, Macmillan, Hampshire, pp.125-153.

Bartlett, W. and Le Grand, J. (1993), 'The theory of quasi-markets', in Le Grand J. and Bartlett, W. (eds), *Quasi-markets and Social Policy*, Macmillan, Hampshire, pp.13-34.

Bartlett, W., Propper, C., Wilson, D. and Le Grand, J. (1994), *Quasi-markets in the Welfare State: The Emerging Findings*, SAUS Publications, Bristol.

Bartlett, W., Le Grand, J. and Propper, C. (1994), 'Where next?', in Bartlett, W., Propper, C., Wilson, D. and Le Grand, J. (eds), *Quasi-markets in the Welfare State: The Emerging Findings*, SAUS Publications, Bristol, pp.269-282.

Bastow, B.W. (1991), *A study of the factors affecting parental choice of secondary schools*, PhD Thesis, London Institute of Education.

Berghman, J. (1996), 'Social exclusion in Europe: Policy context and analytical framework', in Room, G. (ed), *Beyond The Threshold: The Movement and Analysis of Social Exclusion*, The Policy Press, Bristol.

Blake, M. and Openshaw, S. (1995), *GB Profiler '91*, School of Geography, University of Leeds, Leeds.

Blanc, M. (1998), 'Social integration and exclusion in France: Some introductory remarks from a social transaction perspective', *Housing Studies*, vol. 13, 6, pp.781-792.

Bondi, L. (1988), 'The contemporary context of educational provision', in Bondi, L. and Matthews, M.H. (eds), *Education and Society*, Routledge, London.

Bondi, L. and Matthews, M.H. (1988), *Education and Society*, Routledge, London.

Bourdieu, P. (1986), 'The forms of capital', in Richardson, J.G. (ed), *Handbook of Theory and Research for the Sociology of Education*, Greenwood Press, London.

Bourdieu, P. and Passeron, J. (1977), *Reproduction: In Education, Society and Culture*, Sage, London.

Bowe, R., Ball, S.J. and Gold, A. (1992), *Reforming Education and Changing Schools*, Routledge, London.

Bowe, R., Gewirtz, S. and Ball, S.J. (1994), 'Captured by the discourse? Issues and concerns in researching 'parental choice'', *British Journal of Sociology of Education*, vol. 15, 1, pp.63-78.

Bradford, M. (1989), 'Educational change in the City', in Herbert, D.T. and Smith, D.M. (eds), *Social Problems and the City* (2nd Edition), Oxford University Press, Oxford, pp.142-158.

Bradford, M. (1990), Education, attainment and the geography of choice', *Geography*, vol. 75, 1, pp.3-16.

Bradford, M. (1993), 'Population change and education: School rolls and rationalisation before and after the 1988 Education Reform Act', in Champion, T. (ed), *Population Matters: The Local Dimension*, Paul Chapman, London, pp.64-82.

Bradford, M. (1995), 'Diversification and division in the English education system: towards a post-Fordist model?', *Environment and Planning A*, vol. 27, pp.1595-1612.

Bradford, M. and Burdett, F. (1989), 'Privatisation, education and the North-South Divide', in Lewis, J. and Townsend, A. (eds), *The North-South Divide: Regional Change in Britain in the 1980s*, Paul Chapman, London, Chapter 8.

Brown, J.M. (1983), 'The structure of motives for moving: A multi-dimensional model of residential mobility', *Environment and Planning A*, vol. 15, pp.1531-1544.

Brown, P. (1990), 'The 'Third Wave': Education and the ideology of parentocracy', *British Journal of Sociology of Education*, vol. 11, pp.65-85.

Brown, P. (1995), 'Cultural capital and social exclusion: some observations on recent trends in education, employment and the labour market', *Work, Employment and Society*, vol. 9, 1, pp.29-51.

Brown, S. (1996), 'Educational change in the United Kingdom: A North-South divide', in Pole, C.J. and Chawla-Duggan, R. (eds), *Reshaping Education in the 1990s: Perspective on Secondary Schooling*, Falmer Press, London, pp.149-163.

Bruce, M. (1968), *The Coming of the Welfare State* (4th Edition), Batsford, London.

Burdett, F. (1988), 'A hard Act to swallow? The geography of education after the great education reform bill', *Geography*, vol. 73, 3, pp.208-215.

Burrows, R. and Loader, B. (1994), *Towards a Post-Fordist Welfare State*, Routledge, London.

Burrows, R. and Marsh, C. (1992), *Consumption and Class*, Macmillan, Hampshire.

Bush. T., Coleman, M. and Glover, D. (1993), *Managing Autonomous Schools*, Paul Chapman, London.

Byrne, D. and Rogers, T. (1996), 'Divided spaces – Divided school: An exploration of the spatial relations of social division', *Sociological Research Online*, vol. 1, 2, <http://www.socresonline.org.uk/socresonline/1/2/3.html>.

Byrne, D.S., Williamson, W. and Fletcher, B.G. (1975), *The Poverty Of Education: A Study In The Politics Of Opportunity*, Martin Robertson, London.

Carrol, S. and Walford, G, (1997), 'Parents' responses to the school quasi-market', *Research Papers in Education*, vol. 12, 1, pp.32-26.

Chitty, C. (1997), 'Privatisation and marketisation', *Oxford Review of Education*, vol. 23, 1, pp.45-62.

Chubb, J.E. and Moe, T.M. (1988), 'Politics, markets, and the organisation of schools', *American Political Science Review*, vol. 82, pp.1065-1087.

Chubb, J.E. and Moe, T.M. (1990), *Politics, Markets, and America's Schools*, The Brookings Institution, Washington.

Chubb, J.E. and Moe, T.M. (1992), *A Lesson in School Reform from Great Britain*, The Brookings Institution, Washington.

Clarke, D.B. and Bradford, M.G. (1998), 'Public and private consumption and the city', *Urban Studies*, vol. 35, 5-6, pp.865-888.

Clarke, G. and Langley, R. (1996), 'A review of the potential of GIS and spatial modelling for planning in the new education market', *Environment and Planning C*, vol. 14, pp.301-323.

Clarke, G., Langley, R. and Cardwell, W. (1998), 'Empirical applications of dynamic spatial interaction models', *Computers, Environment and Urban Systems*, vol. 22, 2, pp.157-184.

Clarke, J. (1996), 'Public nightmares and communitarian dreams: The crisis of the social in social welfare', in Edgell, S., Hetherington, K. and Warde, A. (eds), *Consumption Matters*, Blackwell, Oxford.

Coates, B.E. and Rawstron, E.M. (1971), *Regional Variations in Britain: Studies in Economic and Social Geography*, Batsford, London.

Conway, S. (1997), 'The reproduction of exclusion and disadvantage: Symbolic evidence and social class inequalities in 'parental choice' of secondary education', *Sociological Research Online*, vol. 2, 4, <http://www.socresonline.org.uk/socresonline/2/4/4.html>.

Cookson, P. (1994), *School Choice*, Yale University Press, Yale.

Coons, J.E. and Sugarman, S. (1978), *Education By Choice: The Case For Family Control*, University of California Press, Berkeley.

Cutler, T. and Waine, B. (1997), *Managing the Welfare State* (2nd Edition), BERG, Oxford.

Daily Telegraph (1996), *Tricks to get into the right school*, 21st November 1996, p.14.

David, M. (1989), 'Education', in McCarthy, M. (ed), *The New Politics of Welfare: An Agenda for the 1990s?*, Macmillan, Hampshire, pp.154-177.

David, M. (1997), 'Diversity, choice and gender', *Oxford Review of Education*, vol. 23, 1, pp.77-87.

Davies, S. (1991), 'Towards the remoralisation of society', in Loney, M., Bocock, R., Clarke, J., Cochrane, A., Graham, P. and Wilson, M. (eds), *The State or the Market*, Sage Publications, London, pp.242-258.

Dennison, W.F. (1983), 'Reconciling the irreconcilable: declining secondary school rolls and the organisation of the system', *Oxford Review of Education*, vol. 9, 2, pp.79-89.

Department for Education (1988), *Public Expenditure White Paper*, HMSO, London.

Department for Education (1992), *Choice and Diversity*, HMSO, London.

Desbarets, J.M. (1983), 'Constrained choice and migration', *Geografiska Annaler*, vol. 65, B, pp.11-21.

Dicken, P. and Lloyd, P.E. (1981), *Modern Western Society*, Paul Chapman, London.

Dore, C. and Flowerdew, R. (1981), 'Allocation procedures and the social composition of comprehensive schools', *Manchester Geographer*, vol. 2, 1, pp.47-55.

Dowding, K. and Dunleavy, P. (1996), 'Production, disbursement and consumption: the modes and modalities of goods and services', in Edgell, S., Hetherington, K. and Warde, A. (eds), *Consumption Matters: the Production and Experience of Consumption*, Blackwell, Oxford, pp.36-65.

Dronkers, J. (1995), 'The existence of parental choice in The Netherlands', *Educational Policy*, vol. 9, 3, pp.227-243.

Dunsire, A. (1990), 'The public-private debate: Some UK evidence', *International Review of Administrative Sciences*, vol. 56, 1, pp.29-62.

Edwards,T. and Whitty, G. (1997), 'Specialisation and selection in secondary education', *Oxford Review of Education*, vol. 23, 1, pp.5-15.

Fitz, J., Edwards, T. and Whitty, G. (1986), 'Beneficiaries, benefits and costs: An investigation of the Assisted Places Scheme', *Research Papers in Education*, vol. 1, pp.169-193.

Fitz, J., Halpin, D. and Power, S. (1993), *Grant Maintained Schools: Education in the Market Place*, Kogan Page, London.

Forrest, K. (1996), 'Catchment 22: New admission system for schools are not making parents happy', *Education*, 8th March, 8.

Gamble, A. (1991), 'The weakening of social democracy', in Loney, M., Bocock, R., Clarke, J., Cochrane, A., Graham, P. and Wilson, M. (eds), *The State or the Market*, Sage Publications, London, pp.259-272.

Gewirtz, S., Ball, S.J. and Bowe, R. (1994), 'Parents, privilege and the education marketplace', *Research Papers in Education*, vol. 9, 1, pp.3-29.

Glass, D.V. (1954), *Social Mobility in Britain*, Routledge, London.

Glatter, R. (1989), *Educational Institutions and their Environments: Managing the Boundaries*, Open University Press, Milton Keynes.

Glatter, R. and Woods, P. (1994), 'The impact of competition and choice on parents and schooling', in Bartlett, W., Propper, C., Wilson, D. and Le Grand, J. (eds), *Quasi-markets in the Welfare State: The Emerging Findings*, SAUS Publications, Bristol.

Glatter, R., Woods, P.A. and Bagley, C. (1997a), 'Diversity, differentiation and hierarchy: School choice and parental preferences', in Glatter, R., Woods, P.A. and Bagley, C. (eds), *Choice and Diversity in Schooling: Perspectives and Prospects*, Routledge, London.

Glatter, R., Woods, P.A. and Bagley, C. (1997b), *Choice and Diversity in Schooling: Perspectives and Prospects*, Routledge, London.

Glover, D. (1992), 'Community perceptions of the strengths of individual schools: the basis of 'judgement'', *Educational Management and Administration*, vol. 20, 4, pp.223-230.

Goldthorpe, J.H. (1980), *Social Mobility and Class Structure* (1st Edition), Oxford University Press, Oxford.

Goldthorpe, J.H. (1987), *Social Mobility and Class Structure* (2nd Edition), Oxford University Press, Oxford.

Golledge, R. G. and Stimson, R. J. (1997), *Spatial Behaviour*, Guilford, New York.

Gorard, S. (1996), 'Three steps to 'heaven'? The family and school choice', *Educational Review*, vol. 48, 3, pp.237-252.

Gorard, S. (1997) *School Choice in an Established Market*, Ashgate, Hampshire.

Gorard, S. (1998), 'School movement in underdeveloped markets: An apparent contradiction', *Educational Review*, vol. 50, 3, pp.249-25.

Gorard, S. and Fitz, J. (1998), 'The more things change... The missing impact of marketisation?', *British Journal of Sociology of Education*, vol. 19, 3, pp.365-376.

Gorard, S. and Fitz, J. (2000), 'Markets and stratification: A view from England and Wales', *Educational Policy*, vol. 14, 3, pp.405-428.

Gore, C. (1995), 'Introduction: markets, citizenship and social exclusion', in Rodgers, G., Gore, C. and Figueiredo, J.B. (eds), *Social Exclusion: Rhetoric, Reality, Responses*, International Labour Organisation, Geneva, pp.1-40.

Haas, L. (1980), 'Role-sharing couples: A study of egalitarian marriages', *Family Relations*, vol. 29, pp.289-296.

Halsey, A.H. (1995), *Change in British Society* (4th Edition), Oxford University Press, Oxford.

Halsey, A.H., Heath, A.F. and Ridge, J.M. (1980), *Origins and Destinations: Family, Class and Education in Modern Britain*, Clarendon Press, Oxford.

Halsey, A.H. Lauder, H. Brown, P. and Wells, A.S. (1997), *Education: Culture, Economy, Society*, Oxford University Press, Oxford.

Harvey, D. (1989), *The Conditions of Postmodernity*, Basil Blackwell, Oxford.

Hay, A.M. (1995), 'Concepts of equity, fairness and justice in geographical studies', *Transactions of the Institute of British Geographers*, vol. 20, pp.500-508.

Hayek, F. (1960), *The Constitution of Liberty*, University of Chicago Press, Chicago.

Higgs, G., Webster, C.J. and White, S.D. (1997), 'The use of geographical information systems in assessing spatial and socio-economic impacts of parental choice', *Research Papers in Education*, vol. 12, 1, pp.27-48.

Hudson, B. (1992), 'Quasi-markets in Health and Social Care in Britain: Can the public sector respond?', *Policy and Politics*, vol. 20, 2, pp.131-142.

Hudson, R. and Williams, A.M. (1995), *Divided Britain* (2nd Edition), Wiley, West Sussex.

Hughes, A. (1999) 'Constructing competitive spaces on the corporate practice of British retailer-supplier relationships', *Environment and Planning A*, vol. 31, pp.819-839.

Jessop, B. (1994), 'The transition to post-Fordism and the Schumpeterian workfare state', in Burrows, R. and Loader, B. (eds), *Towards a Post-Fordist Welfare State*, Routledge, London.

Johnson, D. (1990), *Parental Choice in Education*, Unwin Hyman, London.

Johnson, N. (1991), 'The break-up of consensus: competitive politics in a declining economy', in Loney, M., Bocock, R., Clarke, J., Cochrane,

A., Graham, P. and Wilson, M. (eds), *The State or the Market*, Sage Publications, London, pp.214-230.

Jowett, S. (1995), *Allocating Secondary School Places: Policy and Practice*, NFER, Berkshire.

Judson, D.H. (1990), 'Human migration decision-making: A formal model', *Behavioural Science*, vol. 35, pp.281-289.

Kerckhoff, A.C., Fogelman, K. and Manlove, J. (1997), 'Staying ahead: The middle class and school reform in England and Wales', *Sociology of Education*, vol. 70, 1, pp.19-35.

Kirby, A. (1992), *The Politics of Location*, Methuen, London.

Kitchin, R. and Tate, N.J. (2000), *Conducting research into Human Geography*, Pearson Education, Essex.

Knox, P. (1995), *Urban Social Geography: An Introduction* (3rd Edition), Longman, Essex.

Lauder, H. (1991), 'Education, democracy, and the economy', *British Journal of Sociology of Education*, vol. 12, pp.417-431.

Lauder, H. and Hughes, D. (1999) *Trading in Futures: Why Markets Don't Work*, Open University Press, Buckingham.

Le Grand, J. (1991), *Equity and Choice*, HarperCollins, London.

Le Grand, J. and Bartlett, W. (1993), *Quasi-markets and Social Policy*, Macmillan, Hampshire.

Le Grand, J. and Robinson, R. (1984a), *The Economics of Social Problems: The Market Versus The State* (2nd Edition), Macmillan, London.

Le Grand, J. and Robinson, R. (1984b), *Privatisation and the Welfare State*, Unwin Hyman, London.

Lee, V.E., Croninger, R.G. and Smith, J.B. (1994), 'Parental choice of schools and social stratification in education: The paradox of Detroit', *Educational Evaluation and Policy Analysis*, vol. 16, 4, pp.434-457.

Levacic, R. (1994), 'Evaluating the performance of quasi-markets in education', in Bartlett, W., Propper, C., Wilson, D. and Le Grand, J. (eds), *Quasi-markets in the Welfare State: The Emerging Findings*, SAUS Publications, Bristol, pp.35-55.

Levacic, R. (1995), *Local Management of Schools: Analysis and Practice*, Open University Press, Milton Keynes.

Lidström, A. (1999), 'Local school choice policies in Sweden', *Scandinavian Political Studies*, vol. 22, 2, pp.137-156.

Loney, M., Bocock, R., Clarke, J., Cochrane, A., Graham, P. and Wilson, M. (1991), *The State or the Market*, Sage Publications, London.

Mann, J. (1989), 'Institutions and their Local Education Authority', in Glatter, R. (ed), *Educational Institutions and their Environments:*

Managing the Boundaries, Open University Press, Milton Keynes, pp.117-130.

Manzer, R. (1994), *Public Schools and Political Ideas*, University of Toronto Press, Toronto.

Marren, E. and Levacic, R. (1994), 'Senior management, classroom teacher and governor responses to local management of schools', *Education Management and Administration*, vol. 22, 1, pp.39-58.

Marsden, T., Harrison, M. and Flynn, A. (1998) 'Creating competitive spaces: exploring the social and political maintenance of retail power', *Environment and Planning A*, vol.30, pp.481-498.

Marsden, W.E. (1986), 'Education', in Langton, J. and Morris, R.J. (eds), *Atlas of Industrialising Britain 1780-1914*, Methuen, London, Chapter 29.

Marsh, A. and Mullins, D. (1998), 'The social exclusion perspective and housing studies: origins, applications and limitations', *Housing Studies*, vol. 13, 6, pp.749-759.

Marshall, G. (1997), *Repositioning Class*, Sage, London.

Marshall, G. and Swift, A. (1997), 'Social class and social justice', in Marshall, G. (ed), *Repositioning Class*, Sage, London, pp.178-198.

Martin, C. (1996), 'French review article: The debate in France over 'social exclusion'', *Social Policy and Administration*, vol. 30, 4, pp.382-392.

Martin, D. (1996), *Geographic Information Systems: Socio-economic Applications*, Routledge, London.

Mayet, G. (1997), 'Admissions to schools: A study of Local Education Authorities', in Glatter, R., Woods, P.A. and Bagley, C. (eds), *Choice and Diversity in Schooling: Perspectives and Prospects*, Routledge, London, pp.166-177.

McCarthy, M. (1989), *The New Politics of Welfare: An Agenda for the 1990s?*, Macmillan, Hampshire.

Menahem, G., Spiro, S.E., Goldring, E. and Shapira, R. (1993), 'Parental choice and residential segregation', *Urban Education*, vol. 28, 1, pp.30-48.

Moe, T. (1994), 'The British battle for choice', in Billingsley, K.L. (ed), *Voices On Choice: The Education Reform Debate*, Pacific Institute for Public Policy, San Francisco, pp.23-33.

Morris, R. (1993), *Choice of School: A Survey 1992-93*, Association of Metropolitan Authorities, London.

Mulford, B. (1996), 'Privatisation, the next wave of educational reform: worth the ride?', *International Studies in Educational Administration*, vol. 24, 1, pp.57-66.

Nystuen, J.N. and Dacey, M.F. (1961), 'A graph theory interpretation of nodal regions', *Regional Science Association, Papers and Proceedings*, vol. 7, pp.29-42.

Openshaw, S. and Wymer, C. (1995), 'Classifying and regionalising census data', in Openshaw, S. (ed), *The Census Users Handbook*, Longman, London, pp.239-270.

Papadakis, E. and Taylor-Gooby, P. (1987), *The Private Provision of Public Welfare*, Wheatsheaf Books, Sussex.

Peston, M. (1984), 'Privatisation of education', in Le Grand, J. and Robinson, R. (eds), *Privatisation and the Welfare State*, Unwin Hyman, London, pp.146-159.

Pierson, P. (1994), *Dismantling The Welfare State? Reagan, Thatcher and the Politics of Retrenchment*, Cambridge University Press, Cambridge.

Pinch, S. (1997), *Worlds of Welfare: Understanding the Changing Geographies of Social Welfare Provision*, Routledge, London.

Pole, C. and Chawla-Duggan, R. (1996), *Reshaping Education in the 1990s: Perspectives On Secondary Schooling*, Falmer Press, London.

Pring, R. (1987), 'Privatisation in education', *Journal of Education policy*, vol. 2, pp.289-299.

Raab, G. and Adler, M. (1988), 'A tale of two cities: The impact of parental choice on admissions to primary schools in Edinburgh and Dundee', in Bondi, L. and Matthews, M.H. (eds), *Education and Society*, Routledge, London, pp.113-147.

Ranson, S. (1997), 'From 1944 to 1988: education, citizenship and democracy', *Local Government Studies*, vol. 14, pp.1-19.

Raper, J.F., Rhind, D.W. and Shepherd, J.W. (1992), *Postcodes: The New Geography*, Longman, Harlow.

Rawls, J. (1972), *A Theory of Justice*, Oxford University Press, Oxford.

Reay, D. and Ball, S.J. (1997), 'Spoilt for choice: The working-classes and educational markets', *Oxford Review of Education*, vol. 23, 1, pp.89-101.

Reay, D. and Ball, S.J. (1998), ''Making their minds up': Family dynamics of school choice', *British Educational Research Journal*, vol. 24, 2, pp.431-448.

Rodgers, G. (1995), 'What is special about a 'social' exclusion' approach?', in Rodgers, G., Gore, C. and Figueiredo, J.B. (eds), *Social Exclusion: Rhetoric, Reality Responses*, International Labour Organisation, Geneva, pp.43-55.

Rodgers, G., Gore, C. and Figueiredo, J.B. (1995), *Social Exclusion: Rhetoric, Reality Responses*, International Labour Organisation, Geneva.

Rooke, K. (1993), *To consider the influences, if any, of the marketing techniques upon parental choice at the lower/middle school interface*, MA Professional Studies in Education, University of Leicester, Leicester.

Room, G. (1995), *Beyond The Threshold: The Measurement and Analysis of Social Exclusion*, The Policy Press, Bristol.

Saunders, P. (1990), *A Nation of Home Owners*, Unwin Hyman, London.

Schmida, M. and Katz, Y.J. (1994), 'Parental consideration when choosing high-schools for their children', *Social Behaviour and personality*, vol. 22, 4, pp.337-342.

Seavers, J.C. (1999), *Joint Decision-Making and Rural Mobility in a Rural Environment*, Unpublished PhD Thesis, University of Leicester.

Seldon, A. (1986), 'The riddle of the voucher: An enquiry into the obstacles to introducing choice and competition in state schools', *Hobart Paperbacks*, 21, Institute of Economic Affairs, London.

Sheth, J.N. (1974), 'A theory of family buying decisions', in Sheth, J.N. (ed), *Models of Buying Behaviour*, Harper and Row, New York, pp.17-33.

Shucksmith, M. and Chapman, P. (1998), 'Rural development and social exclusion', *Sociologia Ruralis*, vol. 38, 2, pp.225-242.

Silver, H. (1995), 'Reconceptualising social disadvantage: Three paradigms of social exclusion', in Rodgers, G., Gore, C. and Figueiredo, J.B. (eds), *Social Exclusion: Rhetoric, Reality Responses*, International Labour Organisation, Geneva, pp.57-80.

Smith, D.M. (1994), *Geography and Social Justice*, Blackwell, Oxford.

Smith, G., Smith, T. and Wright, G. (1997), 'Poverty and schooling: Choice, diversity and division?', in Walker, A. and Walker, C. (eds), *Britain Divided: The Growth of Social Exclusion in the 1980s and 1990s*, Child Poverty Action Group, London, pp.123-139.

Somerville, P. (1998), 'Explanations of social exclusion: Where does housing fit in?' *Housing Studies*, vol. 13, 6, pp.761-780.

Spring, J. (1982), 'Dare educators build a new school system?', in Manley-Casimir, M. (ed), *Family Choice in Schooling*, Lexington, Toronto, p.33.

Stillman, A. (1990), 'Legislating for choice', in Flude, M. and Hammer, M. (eds), *The Education Reform Act, 1988*, The Falmer Press, London.

Taylor, C. (2000), *The Geography of the 'New' Secondary Education Market and School Choice in England & Wales*, Leicester University, Leicester (Ph.D thesis).

Taylor, C., Gorard, S. and Fitz, J. (2000) 'A re-examination of segregation indices in terms of compositional invariance', *Social Research Update*, 29.

Taylor-Gooby, P. (1998), *Choice and Public Policy: The Limits to Welfare Markets*, Macmillan, Hampshire.

Teelken, C. (1999) 'Market mechanisms in education: school choice in The Netherlands, England and Scotland in a comparative perspective', *Comparative Education*, vol. 35, 3, pp.283-302.

The Independent (1997a), *The Great School Lottery*, 5th February 1997, p.14.

The Independent (1997b), *School squeeze leaves children without places*, 2nd March 1997, p.15.

The Independent on Sunday (1997), *The admissions nightmare*, Which School? 16th November 1997, p.7.

The Sunday Telegraph (1996), *Bending the school rules*, 8th September 1996, p.18.

Thomas, R.W. and Robson, B.T. (1984), 'The impact of falling school-rolls on the assignment of primary schoolchildren to secondary schools in Manchester, 1980-1985', *Environment and Planning A*, vol. 16, pp.339-356.

Tomlinson, S. (1997), 'Diversity, choice and ethnicity: The effects of educational markets on ethnic minorities', *Oxford Review of Education*, 23, 1, pp.63-76.

Tooley, J. (1997), 'Choice and diversity in education: A defence', *Oxford Review of Education*, vol. 23, 1, pp.103-116.

Turner, J. (1992), *Parental choice of secondary schools*, MA Professional Studies in Education, University of Leicester, Leicester.

Van Der Smagt, T. and Lucardie, L. (1991), 'Decision-making under not-well-defined conditions: From data processing to logical modelling', *Tijdschrift voor Econmische En Sociale Geografie*, vol. 82, 4, pp.295-304.

Vandenberghe, V. (1999), 'Combining market and bureaucratic control in education: an answer to market and bureaucratic failure?', *Comparative Education*, vol. 35, 3, pp.271-282.

Walmsley, D.J. and Lewis, G.J. (1993), *People and Environment: Behavioural Approaches to Human Geography* (2nd Edition), Longman, London,

Waslander, S. and Thrupp, M. (1995), 'Choice, competition, and segregation: An empirical analysis of a New Zealand secondary school market, 1990-1993', *Journal of Education Policy*, vol. 10, pp.1-26.

Weekend Telegraph (1996), *Parents who move house to be top of the form*, 7th February 1996, p.16.

West, A. (1992), 'Factors affecting choice for schools for middle-class parents: Implications for marketing', *Educational Management and Administration*, vol. 20, 4, pp.212-221.

Whitty, G. (1997), 'Creating quasi-markets in education: A review of recent research on parental choice and school autonomy in three countries', *Review of Research in Education*, vol. 22, pp.3-47.

Whitty, G. and Edwards, T. (1998), 'School choice policies in England and the United States: An exploration of their origins and significance', *Comparative Education*, vol. 34, 2, pp.211-227.

Whitty, G. and Power, S. (1997), 'Quasi-markets and curriculum control: Making sense of recent education reform in England and Wales', *Educational Administration Quarterly*, vol. 33, 2, pp.219-240.

Wilkins, C. (1994), *Parental choice: Fact or fiction?*, MA Professional Studies in Education, University of Leicester, Leicester.

Williamson, W. and Byrne, D.S. (1979), 'Educational disadvantage in an urban setting', in Herbert, D.T. and Smith, D.M. (eds), *Social Problems and the City*, Oxford University Press, Oxford, pp.186-200.

Willms, J.D. and Echols, F. (1992), 'Alert and inert clients: The Scottish experience of parental choice of schools', *Economics of Education Reviews*, vol. 11, 4, pp.339-350.

Wilson, P.R. and Elliott, D.J. (1987), 'An evaluation of the Postcode Address File as a sampling frame and its use within OPCS', *Journal of the Royal Statistical Society A*, vol. 150, pp.230-240.

Woods, P. (1992), 'Empowerment through choice? Towards an understanding of parental choice and school responsiveness', *Educational Management and Administration*, vol. 20, 4, pp.204-211.

Index

.

Printed and bound by CPI Group (UK) Ltd, Croydon, CR0 4YY

22/10/2024

01777620-0009